When rearranging the deckchairs, don't forget to look out for icebergs! Functional safety is just one aspect of process safety management.

# Functional Safety In Practice

Harvey T. Dearden

Third Edition

A **SIS**Suite Publication

First published 2016
Second edition published 2018
Third edition published 2020

Copyright © 2020 Harvey T. Dearden

Harvey T. Dearden asserts his moral right
to be identified as the author of this work.

All rights reserved.

No part of this publication may be reproduced, stored in a retrieval system, or transmitted, in any form or by any means, electronic, mechanical, photocopying, recording or otherwise, without the prior permission of the author.

ISBN: 9781696002820

Printed and bound by CreateSpace,
a DBA of On-Demand Publishing, LLC

## DEDICATION

For all those clients that were good enough to
allow me to practice what I preach.

No mere whimsy this; I am sincerely grateful to the many clients who have kept faith with me through their ongoing journeys in functional safety. It is only through working with them on their 'real world' implementation challenges that this book comes to be.

The difference between science and engineering?

  Science is about knowledge, understanding and rigour.

  Engineering is about what works.

Rigour is an exacting muse, but she is not without her austere charms.  Be inspired by her if you will, but be careful not to be enslaved!

# CONTENTS

|    | Abbreviations | x |
|----|---|---|
|    | Preface | xi |
|    | Acknowledgements | xiii |
|    | Introduction | 1 |
|    | Prologue | 4 |
| 1  | Requirements for Compliance | 8 |
| 2  | Functional Safety & Engineering Judgement | 12 |
| 3  | The Safety Lifecycle & Project Execution | 20 |
| 4  | Contractual Provisions | 24 |
| 5  | Identification of Functions | 28 |
| 6  | Demand Modes | 31 |
| 7  | SIL Determination | 34 |
| 8  | Simplified LOPA | 41 |
| 9  | Evaluating SIF Action Failures | 46 |
| 10 | Risk Assessment Tools | 49 |
| 11 | Competence | 54 |
| 12 | Operation & Maintenance | 60 |
| 13 | Certification & Prior-Use | 64 |
| 14 | Systematic Failures & Capability | 75 |
| 15 | Legacy Plant | 84 |

| 16 | Low Integrity Functions | 90 |
|---|---|---|
| 17 | Verification & Validation | 95 |
| 18 | Functional Safety Assessments (& Audits) | 97 |
| 19 | Element Metrics: Useful Life, Mission Time, MTBF, MTTF, MRT, MTTR | 101 |
| 20 | Adjusting Equipment Failure Rates for Duty | 104 |
| 21 | Validating Failure Rates and Managing End of Useful Life | 114 |
| 22 | Sharing of Elements Between BPCS & SIS | 125 |
| 23 | Trip Setting Nomination and Process Safety Time | 130 |
| 24 | Comparison Alarms | 137 |
| 25 | Proof Testing | 144 |
| 26 | Regulatory Overview | 159 |
| 27 | Functional Safety Management Planning | 163 |
| 28 | Aggregate Risk & Risk Profiles | 167 |
| 29 | Evaluation of Compound SIF | 171 |
| 30 | Leading Indicators and FSA4 | 177 |
| 31 | Mitigation Systems | 183 |
| 32 | Modification, Decommissioning and FSA5 | 189 |
| 33 | The Suspended Load Process Safety Model | 191 |
| 34 | Security: Physical & Cyber | 195 |
| 35 | SIL & Cybersecurity Levels (SL) | 198 |
| 36 | Common Cause & Beta Factors | 204 |

| 37 | Proof Test Coverage Nomination | 210 |
| 38 | Multiple SIF Layers | 216 |
| 39 | Human Error | 220 |
| 40 | Overrides & Resets | 224 |
| 41 | Consequence Mitigation in LOPA | 228 |
| 42 | SIL4 | 231 |

| | Errata | 233 |
| | Index | 234 |
| | **SIS**Suite™ | 237 |
| | About the Author | 238 |

# ABBREVIATIONS

| | |
|---|---|
| ALARP | As Low As Reasonably Practicable |
| BPCS | Basic Process Control System |
| C&E | Cause and Effect |
| CPD | Continuing Professional Development |
| DCS | Distributed Control System |
| EA | Environmental Agency |
| E/E/PE | Electrical/Electronic/Programmable Electronic |
| ESD | Emergency Shutdown |
| FE | Final Element |
| FMEA | Failure Mode Effects Analysis |
| FPL | Fixed Programmable Language |
| FSA | Functional Safety Assessment |
| FSMP | Functional Safety Management Plan |
| FVL | Full Variability Language |
| HFT | Hardware Fault Tolerance |
| HSE | Health and Safety Executive |
| LOPA | Layer of Protection Analysis |
| LVL | Limited Variability Language |
| MRT | Mean Repair Time |
| MTBF | Mean Time Between Failures |
| MTTF | Mean Time To Failure |
| MTTR | Mean Time To Restore |
| O&M | Operation & Maintenance |
| PFD | Probability Failure on Demand |
| PFH | Probability Failure per Hour |
| PLC | Programmable Logic Controller |
| PIU | Proven-In-Use |
| PU | Prior-Use |
| QA | Quality Assurance |
| QRA | Quantitative Risk Analysis |
| RRF | Risk Reduction Factor |
| SC | Systematic Capability |
| SFF | Safe Failure Fraction |
| SIF | Safety Instrumented Function |
| SIL | Safety Integrity Level |
| SIS | Safety Instrumented System |
| SRS | Safety Requirements Specification |

# PREFACE

This book is not intended as a primer on functional safety; there is plenty of other material available that will fulfil that purpose. (I did toy with calling it 'Every Engineer's SECOND Book of Functional Safety'.) This book is intended to help bridge the gap between the theory of the standards and their practical implementation, and it does assume a working knowledge of the functional safety standards. Some points are repeated in different chapters because they are relevant in more than one context and the intention has been to make each individual discussion largely stand alone.

**For this third edition**, I have again adopted the expedient of adding the new material as Chapters (34-42) that follow those established in the previous edition (with some minor corrections and clarifications). The aim remains to bring a practicable perspective to the implementation of functional safety. Purists may point to a want of rigour in places, but that need not unduly concern those with a job to do rather than talk about; the broad uncertainties remain and many aspects are necessarily treated qualitatively rather than quantitively. This is particularly true of cybersecurity concerns and the advent of the related set of standards. I do not attempt a comprehensive treatment of these issues, but hopefully the discussion I offer will help place them in their proper perspective in the context of functional safety. I continue to advocate the need for appropriately informed engineering judgement[a] in matters of functional safety (amongst others).

---

[a] Those with an interest in the underpinning philosophy might enjoy my book *'Professional Engineering Practice: Reflections on the Role of the Professional Engineer'*, 2$^{nd}$ Ed., J.R. Smith Publishing, 2017.

The considerations outlined in the book have also informed the development of a suite of software tools (**SIS**Suite™) in support of functional safety management, which is available from SISSuite Ltd.

Harvey T. Dearden
North Wales
2020

## ACKNOWLEDGMENTS

Besides the clients referred to in my dedication, particular thanks are due to Mr Nic Daniels (of LyondellBasell) for the many catalysing discussions we have had; Mr Peter Morriss (formerly of Shell UK and now an independent consultant) for his moral support throughout; and Mr Martin Hold (of the HTS Engineering Group) who has shared my vision in creating the SISSuite™ toolset.

For this 3rd Edition, I wish also to acknowledge the support of Mr Mirek Generowicz of I&E Systems Pty Ltd., who was good enough to review most of the new material for me. Also that of Dr Andy Brazier of AB Risk Ltd. for his review of the chapter on Human Factors.

I lay exclusive claim to any residual errors.

# INTRODUCTION:
# (FUNCTIONAL SAFETY) GAMES THAT PEOPLE PLAY

The IEC 61508 standard comes in 7 parts with a total of approaching 600 pages; it cannot be described as an 'easy read'. The core provisions are relatively straightforward, but there are many subtleties and qualifications that can obscure the path to righteousness. There is plenty of scope for confusion, and a common mistake (or piece of trickery when wilful) is to take individual clauses or provisions and quote them out of context; they remain 'true', but their significance or relevance may be misrepresented.

Human beings, with all their frailties, will bring their individual political agendas to bear. There is however only one properly legitimate agenda; the reduction of risk to be 'As Low As Reasonably Practicable' (ALARP). Functional safety is one (and certainly not the only) means of delivering this.

This book is not called 'Functional Safety in Theory'; there is no shortage of material that expounds the theory of functional safety, but precious little about the practice i.e. actual implementation in what is sometimes called the 'real world', where we routinely meet a variety of constraints that do not allow the theoretical model to be fully realised. This book offers guidance on practical implementation; in this it must be immediately acknowledged that there is a spectrum of possibilities. (Be wary of those that tell you there is only one definitive way to skin a dead cat!) Readers seeking definitive answers are likely to be disappointed; what they are offered instead is an outline of the considerations that should inform their choices and judgements. And make no mistake, there are many judgements to be made in the pursuit of functional safety. Absolute compliance is a fantasy that is promoted by the unscrupulous or the misguided. Insistence on absolute and rigorous compliance with the standard would likely see your project very seriously stalled. The standard is so broad in its scope, and so detailed in its provisions, that absolute compliance in every particular is hardly to be believed (or required); it is an ideal that we approach asymptotically. It is easy for an 'expert' to point to a non-compliance, but that need not necessarily be a matter for concern unless the non-compliance can be shown to materially compromise safety; there is a corresponding requirement to exercise professional judgement.

We are rightly encouraged to adopt relevant good practice, and the UK's Health and Safety Executive point to the functional safety standards as the pertinent benchmark. IEC 61508 (or the relevant sector standard 'daughter') is the ruler they will hold up to measure duty holders by. There is however no legal obligation to comply; the legal obligation (in the UK anyway) is to manage risk to be As Low As Reasonably Practicable (ALARP).

Note that the phrase is 'good' practice, not 'best'; 'best' is exceptional. There may well be someone that does it better than you, but that does not necessarily mean you should adopt their practices, since they may not be appropriate in the context in which you are operating. In particular, be wary of the pursuit of rigour for its own sake (or of those that would urge this upon you). Rigour does not sit alongside 'motherhood' and 'apple pie' as a universal good. If it obfuscates or demands additional resource without providing material benefit it (she?)

is not earning its (her[a]) keep.

Those good practices that were established before IEC 61508 was invented have not been rendered obsolete, and duty holders that have historically adopted good practice are likely to be already well up the compliance curve. Do not trust those that would have you believe that you need to 'rip it all out and start again'. You should apply sanity checks to any proposal; 'does this look sensible?' If it doesn't, and it has been offered in good faith, there are two possibilities: either you have not understood, or the proposer hasn't. Which is it? You need to know. I will go further; if you are acting in the capacity of a professional engineer, you are under a professional, moral obligation to find out which it is.

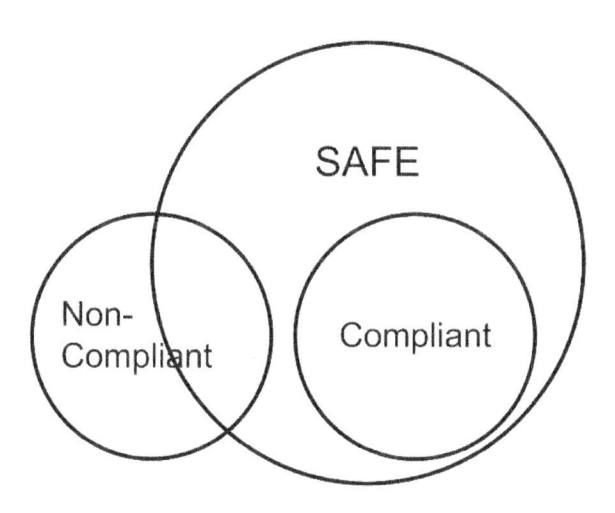

This Venn diagram represents possibly the most important notion in the book! Deviations from absolute compliance may be perfectly acceptable and may represent the optimal solution in the 'real world'. Be wary of those that point to non-compliance without reference to its significance in the context of your project.

---

[a] See page vi

# PROLOGUE: RUSSIAN ROULETTE (PROCESS STYLE)

For those unfamiliar with the term, Russian roulette refers to the loading of a revolver with a single round, spinning the chambers so that the position of the live round is unknown and then placing the muzzle against your temple and pulling the trigger. With just six chambers, the classic version of the game is extremely risky and, understandably, you would likely refuse to play. But what if there were a thousand chambers and you pulled the trigger just once a year;[a] ready to play? Perhaps you already are - with your process plant?

The playing of Russian roulette (process style) was recognised by Diane Vaughan, who, in her study of the Challenger space shuttle disaster [1], coined the phrase 'Normalisation of deviance'; the erosion of the O-rings that led to the disaster was a known occurrence, but NASA were lulled into a false sense of security because the observed erosion had never caused an actual failure. Familiarity, if not actually breeding contempt, had led to a discounting of the risk; the partial failure 'deviance' became accepted as routine, it became 'normalised'.

We must counter this normalisation of deviance by maintaining a sense of vulnerability; near miss recognition is key. Each near miss is a 'click' as the hammer falls on an empty chamber. And, of course, just because your plant has never suffered a major accident in 30 years operation does

---

[a] The UK's Health and Safety Executive (HSE) identify this as the 'intolerable' boundary for individual risk of fatality.

not mean that all the chambers are empty! If an occurrence rate is very low it will likely lie outside our experience and our intuitive feel for the risk is likely to be correspondingly poor. (When I undertake quantitative risk assessment (QRA) work, I often have Richard Feynman metaphorically whispering in my ear *"If a guy tells me the probability of failure is 1 in $10^5$, I know he's full of crap!"*[2] (Investigating the Space Shuttle *Challenger* Disaster). I regard this as healthy; all those engaged with QRA should have a metaphorical Richard Feynman sitting on their shoulders. Typically such extremely low probabilities or frequencies are useful as indications of relative rather than absolute risk.)

There is a lot of talk and study in relation to safety cultures and the issues are well understood. The more challenging aspect is cultivating appropriate management practice. A variety of models are presented, all of which can be said to offer some value, but many of them suffer from over-elaboration through attempts to introduce increased rigour, which is often pursued for its own sake regardless of whether it actually provides net value. If the provisions are not broadly intuitive then their take up is likely to be limited or distorted and the wider safety management may be degraded by a loss of credibility. The moment that a management provision is perceived as a 'going through the motions exercise', there is diminishment.

One cultural obstacle may be the designation of a dedicated safety department; for 'department' we might read 'silo'. Certainly your safety resource should facilitate and provide support, but they should not have exclusive ownership of safety; safety should be 'owned' by the business management and the operations and maintenance functions. Whatever your role, if you discover what you believe to be a significant failing or misunderstanding that may impact on safety, you are, as a matter of personal professional integrity, obliged to bring this to the attention of the appropriate managers and make sure that it is addressed responsibly and not 'swept under the carpet'.

This is potentially fraught territory; we must be prepared to take an unpopular stance. A process engineer of my acquaintance provides an interesting example: the plant was on course to break their annual production record with plaudits all round in prospect, but following a plant modification he recognised that the plant was behaving differently and was concerned that the modification, whilst successful from a production perspective, had introduced an increased risk potential. He refused to sanction continuing operation until enhanced protection was commissioned – the delay caused them to miss the record (but no one died). This takes courage.

The manner in which you raise issues will be important. An intemperate rant or attempt to ridicule transgressors is unlikely to help 'sell' the message. A sympathetic, but sincerely concerned intervention is more likely to get results. If we make ourselves objectionable then we risk being ostracised and losing any ability to influence things for the better. Intervention should not just be on questions of specific equipment failures and operating and maintenance practices, but also in defence of wider safety management and risk assessment practices.

We should acknowledge however that this can be overdone; unwarranted conservatism (the professional engineer's equivalent of 'crying wolf') may undermine our credibility and may lead to the spurious allocation of resources to where they will achieve very little. If there are the equivalent of 100,000 chambers, we cannot reasonably object to the trigger being pulled every 10 years.[b]

In design and risk assessment there may well be full cognizance of hazard consequence, but in day-to-day operation and maintenance, the prevailing sense may become one not of vulnerability, but rather of remote likelihood, and yet likelihood may be growing as barriers are progressively degraded through ageing and obsolescence (of both plant and people) and replacement and changes (in both plant and people) and the Swiss cheese holes grow into increasing alignment.

Although not explicitly identified in the usual models, the integrity of

---

[b] The HSE identify this level of risk as 'broadly acceptable'.

individual engineers is a critical requirement in the establishment and maintenance of the defences against major accident hazards. If this is not in place then all bets are off. We see increasing promotion of accreditation and certification of specific skill sets, but when we examine incident case histories, we typically find not want of knowledge or skills, but rather of diligence. This is a core component of professionalism. If all professional engineers embrace their responsibilities in this regard they can provide mutual support to help drive the culture in the appropriate direction.

If we encounter individuals who, whether unwittingly or no, are propagating errors in process safety management, we are under a professional obligation to intervene; we might think of it as 'normalisation of deviants'! If the transgressions are systemic and wilful, and clearly hazardous, then there is a duty to blow the whistle long and hard. Silence will be construed as acquiescence, and any professional who has acquiesced in a process safety failure cannot claim to have 'just been following orders'.

If you lay claim to the title 'professional engineer' you cannot turn a blind eye or walk by on the other side. 'Not my responsibility' will not wash. If you have reason to suspect that the trigger is being pulled more often than anticipated, or that more chambers are loaded than expected, you have a professional obligation to ensure it is investigated, even if you are not the one pulling the trigger.

### References
1. Vaughan, D. (1996). The Challenger launch decision: risky technology, culture, and deviance at NASA. Chicago: University of Chicago Press.

2. Feynman, Richard P. (2007). "What Do You Care What Other People Think?", Penguin Books.

# 1. REQUIREMENTS FOR COMPLIANCE

So you have identified the safety instrumented functions (SIF) and you have identified the SIL targets, so far so good, but then we need to understand what the requirements are for compliance. We may categorise these as follows:

**Probability of Failure (PFD/PFH)**
There is generally a good appreciation of this requirement; each function must offer a PFD/PFH that sits in the appropriate band for the SIL target that has been nominated. There are some subtleties here however. If the SIL target was identified through a qualitative approach (such as risk graph or risk matrix) then strictly the PFD/PFH target should be at the low end of the SIL band. If a quantitative approach has been used (e.g. LOPA) then the PFD/PFH required to meet the tolerable risk target should become the target for the function design and implementation.

Reference IEC 61508, Part 1, Clause 7.10.2.7 a) Note 1. (buried deep!)
*"The specified value for the target failure measure can be derived using a quantitative method (see 7.5.2.3). Alternatively, when the safety integrity requirement has been developed using a qualitative method and expressed as a safety integrity level, then the target failure measure is derived from Table 2 or 3, as appropriate, according to the safety integrity level. In this case the specified target failure measure is the smallest*

*average probability of failure or failure rate for the safety integrity level, unless a different value has been used to calibrate the method."*

This provision is often overlooked and might well seem too conservative, particularly if a conservative approach has been adopted with the qualitative assessment itself. Clearly, specifying the other end as a target would be a cynical manipulation, but it might be considered reasonable to use a mid-band target. If in all other respects the design was compliant then I would not be troubled if the PFD was at the geometric middle of the relevant SIL band e.g. 0.0316 for SIL1, 0.00316 for SIL2. (The point being that this is a factor $\sqrt{10}$ from either end.)

A variety of techniques are available for calculating PFD/PFH with a spectrum of possibilities in terms of sophistication. There are some intriguing mathematical challenges here, but the increased sophistication of some of these tools is typically of questionable value, since there are such very broad uncertainties in so many of the other aspects of functional safety; element failure rates, demand rates, SIL determination etc. In the face of these uncertainties, the additional 'accuracy' of the more sophisticated calculation tools is not typically a critical concern. For most purposes the simplified equations are perfectly satisfactory.

**Hardware Fault Tolerance (HFT)**
Again, there is generally a good appreciation of this requirement and two 'routes' to compliance are identified in IEC 61508 ($1_H$ and $2_H$). Route $1_H$ identifies the HFT requirement on the basis of the Safe Failure Fraction (SFF) and the type (A/B) of the equipment elements used to build the function. The idea of SFF is to promote robustness in the design by requiring redundancy when SIL targets are higher and/or element safe failure fractions are low. In some quarters this approach is very unpopular; it is true that as a matter of mathematical rigour there are some inconsistencies that can lead to some anomalous positions, but it was conceived as a relatively straightforward means of promoting robustness. Route $2_H$ under IEC 61508 identifies the HFT requirement directly on the basis of the SIL, without reference to the SFF, but places

additional requirements on the rigour with which the equipment failure rates are identified. IEC 61511 2$^{nd}$ Edition now adopts Route $2_H$, but points to Route $1_H$ under IEC 61508 as an alternative.

**Systematic Integrity**
This is less well understood and is sometimes overlooked entirely (!), but is a distinct 'link-in-the-chain' without which there can be no demonstration of functional safety to a SIL. It is essentially a measure of the 'fitness-for-purpose' of the SIF equipment and its installation, and is achieved through the incorporation of appropriate 'Techniques and Measures' in the design and implementation aspects that defend against the introduction of systematic failures (refer Annex A & B of Part 2). The requirement for this systematic integrity has always been present in the standard, but with the 2nd edition it was explicitly identified with a systematic capability (SC) parameter for elements. (See Chapter 14 for a discussion of systematic failures.)

Software brings particular challenges and IEC 61508 Part 3 offers three routes to compliance:

$1_S$: compliant development. Compliance with the requirements of part 3 for the avoidance and control of systematic faults in software

$2_S$: proven-in-use. Provide evidence that the element is proven-in-use. (See 7.4.10 of IEC 61508-2)

$3_S$: assessment of non-compliant development. (In accordance with clause 7.4.2.13 of Part 3, which is a bit like $1_S$ but tailored appropriately with reference to Annex C: Properties for software systematic capability).

**Functional Safety Management**
These aspects may be more nebulous than the three above, but are still required for compliance. They are NOT optional extras! This category includes:

## REQUIREMENTS FOR COMPLIANCE

- Planning (Chapters 3 & 27)
- Project Management (Chapters 3, 4, 11)
- Functional Safety Assessments and Audits (Chapter 18)
- Competency (Chapter 11)
- Operation & Maintenance (Chapter 12)
- Change Control (Chapters 12 & 31)

## 2. FUNCTIONAL SAFETY & ENGINEERING JUDGEMENT

It is the role of a professional engineer, having acquired the appropriate competencies, to exercise professional judgement with due regard to pertinent guidance. In terms of the functional safety standards, engineers should recognise that we approach compliance asymptotically along a curve of diminishing return; we may approach closer and closer to full compliance, but it requires ever increasing effort and investment. There is a point where the marginal increase in compliance does not warrant the additional effort, which may be more gainfully employed on other safety concerns. Professional judgement must be exercised to identify when this point has been reached. If engineers do not exercise appropriate judgements, there is a real danger of people tying themselves up in all sorts of metaphorical knots in an attempt to establish total, rigorous compliance with the standards, with the result that wrongheaded approaches may be adopted and resources and efforts misdirected from where they may well yield a better safety return.

Note that 'exercising judgement' does not mean the same as 'going out on a limb', which would imply accepting a significantly higher degree of risk than would otherwise arise; it will often simply come down to a matter of employing some common sense rather than 'going through the motions' of strict compliance even when there is little or no benefit (or even conceivably negative 'benefit').

The wish of some engineers to avoid any accountability is recognised by some as a marketing opportunity. Occasionally you will find interested parties use 'scare tactics' to promote their product or service. It is all very well picking out individual clauses from the standards to 'demonstrate' a particular requirement, but without proper consideration of the context and the underpinning philosophy of the standard, it is easy to end up with a wrong-headed approach.

If ever you are being told that you need to 'rip it all out and start again', whether metaphorically or literally, you need to take a step back and consider whether the 'logic' that is being deployed has carried you to a place that is no longer sensible. Do not fall into the trap of assuming a logical proposition is necessarily a sensible one. (Which is itself an error in logic!) Careful consideration of the starting point and the explicit and implicit assumptions being made may reveal that a logical conclusion may be far from sensible.

As an example, consider the following: a company had been hired to do a review of existing safety system provisions against the requirements of IEC 61508. A typical argument deployed in their report ran as follows (here distilled to its essence):

A legacy system has a number of components. None of these is certified as being suitable for deployment in a SIL rated system and the safe failure fractions are not identified. Therefore conservatively assume that the safe failure fractions are less than 60%. But the standard specifies a safe failure fraction of less than 60% is 'Not Allowed' for the specified SIL target. Therefore replace all the components.

Logical, but very far from sensible: The equipment had been procured and installed in accordance with good practice and there was no suggestion that it was not fit for purpose. An intelligent, responsible review, exercising appropriate judgement would have identified this.

The important thing is to bring a considered, systematic and responsible approach to these matters; insistence on an entirely rigorous approach where not only the spirit of the standard is met, but also the letter, may well produce unwarranted distortion in deployment of resources, resulting in a net loss of safety. Engineers should recognise that there is often a trade-off between rigour and robustness; a more rigorous approach may look good on paper, but may well prove more fragile in operation, with a tendency to suffer inadvertent corruption and with people prompted to make short cuts. This can in turn undermine the wider safety culture within an operation.

Many organisations have been prompted to generate their own guidance, particularly on matters of SIL determination, with a variety of risk matrix and risk graph offerings. These may well be useful as an aid to assessment, but these tools have limited resolution and their 'calibration' may not always be appropriate for a given context. It would be a mistake to simply accept the output of these tools as definitive. These tools should rather be considered as a means of probing plant design and provisions and identifying potential anomalies where provisions appear inconsistent with the risks. If the output of the tool is at odds with your judgement be prepared to examine the situation more deeply. From where does the apparent discrepancy arise? Is the assessment flawed or should your judgement itself be 'recalibrated' by newly identified considerations? There may well be qualifying circumstances or provisions that have not been properly recognised in the application of the tool. If you are persuaded that there is a real shortfall in safety provisions, then there is a corresponding obligation to address this.

**Uncertainty in Estimation**
The following statement, drawn from the 'Guidance on Risk for the Engineering Profession' published by the Engineering Council [1], highlights this concern:

• *bear in mind that risk assessment should be used as an aid to professional judgement and not as a substitute for it*

Uncertainty is a feature of many aspects of risk management. We should be particularly aware of this in the context of the functional safety standards. In order to identify the appropriate Safety Integrity Level (SIL) and implement appropriate design and test provisions, we need to employ a variety of guesses about, for example; hazard consequences, demand rates, failure rates, and safe failure fractions. Some of these guesses may be your own; some may be imported from colleagues. Some may be sourced from vendors or other third party sources. I use the term 'guesses' for dramatic effect, to bring home the point that functional safety is NOT an exact science. Being engineers, we do not simply guess however, we make considered and, where appropriate, substantiated judgements. It is by this means that our guesses earn the superior designation of 'estimates'. They remain estimates however; and there really is no point troubling over the third significant figure, or even possibly the second. Why should the SILs be separated by a factor 10 in terms of PFD/PFH? Why not a factor 2 or 3? It is an implicit acknowledgement of the inexact nature of functional safety; we look to implement risk reduction with an appropriate order of magnitude. Remember that whereas random failures admit of analysis through statistical calculations, the same cannot be said of systematic failures. A particularly pertinent note in 61511 says "The safety integrity level is defined numerically so as to provide an objective target to compare alternative designs and solutions. However, it is recognised that, given the current state of knowledge, many systematic causes of failure can only be assessed qualitatively."

Do not be fooled by the nature of the standards (in particular their bulk!), the way they are written or the apparent rigour with which the requirements are formulated, into thinking that functional safety is anything like an exact science. The standards are a challenging read and it can be difficult to see the wood for the trees. The real thrust (wood) of the functional safety standards is the requirement that you approach your assessment of protection provisions in a systematic, considered, risk-based manner. Prior to the introduction of the standards, process plant operations were not giving rise to exceptional numbers of fatalities;

established good practice has not been rendered obsolete, although on seeing the scope of the standards you could be forgiven for thinking previous practice must have proved woefully inadequate.

The standards do not discuss the uncertainties of estimates and their propagation through dependent analyses. The implication is that estimates should have a suitably narrow uncertainty at a suitably high confidence level. But the same can be said of any engineering calculation. This is an area that is rightly left to professional judgement. Any attempt to place this on a more rigorous footing would complicate matters enormously without bringing corresponding benefits. From a practical perspective, the uncertainty in estimates employed in functional safety studies can be very broad; we are essentially looking to establish an appropriate order of magnitude of risk reduction. An estimate may typically be out by a factor 2, or 3 or more without critically affecting the analysis. Periodic reviews against operational experience (as called for in the standards) protect against unduly optimistic estimates.

There is a note in the standards that says estimates for failure rate should have a confidence level of at least 70%. The implication here is that there should be a less than 30% chance that the actual failure rate is higher than your estimate. But without a population of estimates and some notion of their distribution this is not very helpful. And simply specifying a confidence limit does not constrain the factor by which our estimate could be exceeded; it only constrains the probability of its being exceeded. None of this need trouble us however. This is the modern equivalent of debating how many angels can dance on the head of a pin. Attempts to introduce such mathematical rigour are failures to recognise the nature of the subject and the real aim of the standard; a considered, systematic, risk based evaluation of protection provisions.

Given the tendency to conservatism in both the assessment of the hazard consequences (and therefore their tolerable interval), and demand rates placed upon protection systems, there is often a corresponding tolerance in under estimation of failure rates. If tolerable interval and demand rate are both conservative by a factor of only 1.7, we may use a failure rate figure that is low by a factor 3.0 (being ≈ $1.7^2$) and still deliver the risk reduction required for the hazard.

**Confidence in Equipment Compliance**
Some will place their faith in component design certificates as a means of gaining assurance, but a certificate is simply a declaration of someone else's estimates. They are one possible means of acquiring confidence that equipment is fit for purpose, but they are not the only way, and not necessarily to be preferred. Certification is very variable in its quality. Some certificates only address the hardware and do not cover any software deployed, which rather limits the usefulness of the certificate to the user (it may help substantiate an estimate of random hardware failure rate). Some may certify the equipment under conditions that are not representative of expected use. A critical review of a certificate (or its underpinning report) may reveal deficiencies in the certification process, particularly in respect of the software. The certified declarations may be a starting point, but may need qualifying in respect of the anticipated deployment of the equipment.

The alternative approach is to use a 'prior-use' argument in accordance with IEC 61511. There are any number of people that will queue up to tell you how hard the prior-use (PU), argument is to deploy. (Most with an interest in certification.) Some shameless scaremongering has been employed. Is it really so difficult? Consider: why would the standard committee include the PU provision and then make it next to impossible for the user to adopt as a practicable option? The compilation of PU substantiation is not necessarily that onerous, particularly for SIL1 and 2.

Some people seem to regard prior-use as the poor relation of design certification, but this is not correct; good experiential evidence is to be preferred to theoretical prediction and is ultimately the only means of validating reliability modelling. FMEA studies are routinely employed in certification of equipment failure rates, but these assessments will be for operation under reference conditions and will not include: scaffold pole strikes, being stepped upon, being leaked upon, process/environment excursions, process connection failures, maintenance failings etc. An estimate based upon established experience, even if not offering the same rigour in methodology, may well be superior to the 'rigorous' theoretical assessment. You may also find that suitably certified equipment is simply not available; your only recourse will be to a prior-use argument.

In terms of a system (as distinct from a component), it is questionable whether you should look to a certificate (or audit report) to gain complete assurance of compliance of a system with a given SIL. The scope of the standards is so broad, the lifecycle so extensive, the detailed provisions so pervasive (including project procedures remember), that total compliance is hardly credible. It is an aspiration that we approach asymptotically. Look for compliance in all essentials. Look for compliance to the point of being fit for purpose. Exercise your professional judgement.

Sometimes it may be appropriate to refine your estimates with Quantified Risk Assessment, using tools such as fault tree studies. But do not kid yourself; they remain estimates, albeit of a higher order. The point is that instead of simply, for example, judging hazard rate say as belonging to a certain band of frequencies (possibly based on some back-of-envelope calculation), you compute the rate for the top hazard event as a function of a larger number of component estimates at the bottom of your fault tree.

**Conclusion**

Before you undertake a program of work to introduce more rigorous compliance, or refine your calculations or enhance the substantiation for your estimates, be careful to consider whether the additional efforts are warranted and whether you will be adding useful value. You may spend a great deal of time and effort agonising over the possible refinements, but the apparent improvement in precision may be illusory, or even if real, may be largely irrelevant in the face of other uncertainties. Recognise that functional safety is not an exact science. Do not look for 'bombproof', absolute assurances of compliance in every particular. Once you have acquired the appropriate competencies, be willing to exercise your professional judgement. It is then possible to meet the spirit of the standards in a relatively straightforward manner. The functional safety standards are so comprehensive in their scope and so detailed in their provisions that absolute compliance in every particular is hardly to be believed, nor necessarily to be required.

**References**

1.  Guidance on Risk for the Engineering Profession, Engineering Council, (March 2011).

## 3. THE SAFETY LIFECYCLE & PROJECT EXECUTION

The Safety Lifecycle is familiar to everyone with a working knowledge of functional safety, but what is not typically recognised is that the Safety Lifecycle as identified in IEC 61508/61511 does not work! The standard itself acknowledges the limitations of the simplified representation it offers; (Part 1, 7.1.1.4 )

*'The overall, E/E/PE system and software safety lifecycle figures are simplified views of reality and as such do not show all the iterations relating to specific phases or between phases. Iteration, however, is an essential and vital part of development though the overall, E/E/PE system and software safety lifecycles.'*

That has not prevented it being used in thousands of presentations and publications without the health warning that should accompany it.

The representation in the standard is of a so-called 'waterfall model' in which the successive phases follow one another in a purely sequential fashion and the information only ever moves 'downstream'. Now, in theory, you could implement a project in exactly this manner, but the consequence would be a massive increase in the project timescale and expenditure. Back in the real world we must acknowledge that a purely serial process is not going to deliver the needed safety instrumented functions in a timely and cost effective manner. The truth is, that as we execute later phases we are likely to identify unanticipated consequences of earlier 'upstream' decisions, or come to realise deficiencies in earlier work. Rather than remain locked in to an unhelpful situation, it makes more sense to consider revising the earlier work.

There is nothing new in this notion of course; this is absolutely routine territory in project engineering; we progress design and delivery in an iterative manner. In the same way, the safety lifecycle, as implemented in practice, is full of iterative loops; if, for example, implementation of a given SIL target starts to look unduly burdensome it would be entirely sensible to revisit the SIL determination and do a sanity check; if this reveals misplaced conservatism, a reduction in the SIL target may be entirely legitimate. If your check confirms the original target you can tackle the engineering challenges anew with confidence that the effort is truly warranted. Recognise also that the project protocols you adopt will themselves also be subject to iterative development.

Note also that in the real world 'parallel processing' will take place. You will not wait for an earlier phase to be fully completed, and checked, approved and signed off, before beginning work on the next downstream phase. (You would not wait for all the HazOp actions to be closed out before advancing the design.) Draft versions of phase deliverables may be usefully issued to allow preliminary development of dependent works.

For these reasons, you should not attempt to map your project execution to the simplified model in the standard on a one-to one basis; that way lies a failed project. That is not to say, however, that the standard model is useless; it IS useful for informing our project execution protocols; it DOES identify the relationship between different phases. We need to ensure that all the key aspects are properly and responsibly addressed, so that there no gaps for things to fall into (e.g. reputations), and so that we end up with an appropriate, coherent and cost effective solution.

There is in fact no one definitive model for project execution; the specifics will depend on resources (capabilities and availability), and organisational structures and cultures. The number and nature of contractual interfaces (e.g. suppliers, contractors, integrators or consultants) should also be taken into account. It is these considerations that should help you establish your functional safety management plan.

The common approach is to generate a high level 'motherhood and apple pie' statement that allows the functional safety management plan 'box' to be ticked, but which in truth adds little real value and is largely regarded as something of a going-through-the-motions exercise. Your plan does not have to be completely locked down at the outset; it too should undergo iterative change and be a 'living' and useful document throughout the project execution.

THE SAFETY LIFECYCLE & PROJECT EXECUTION

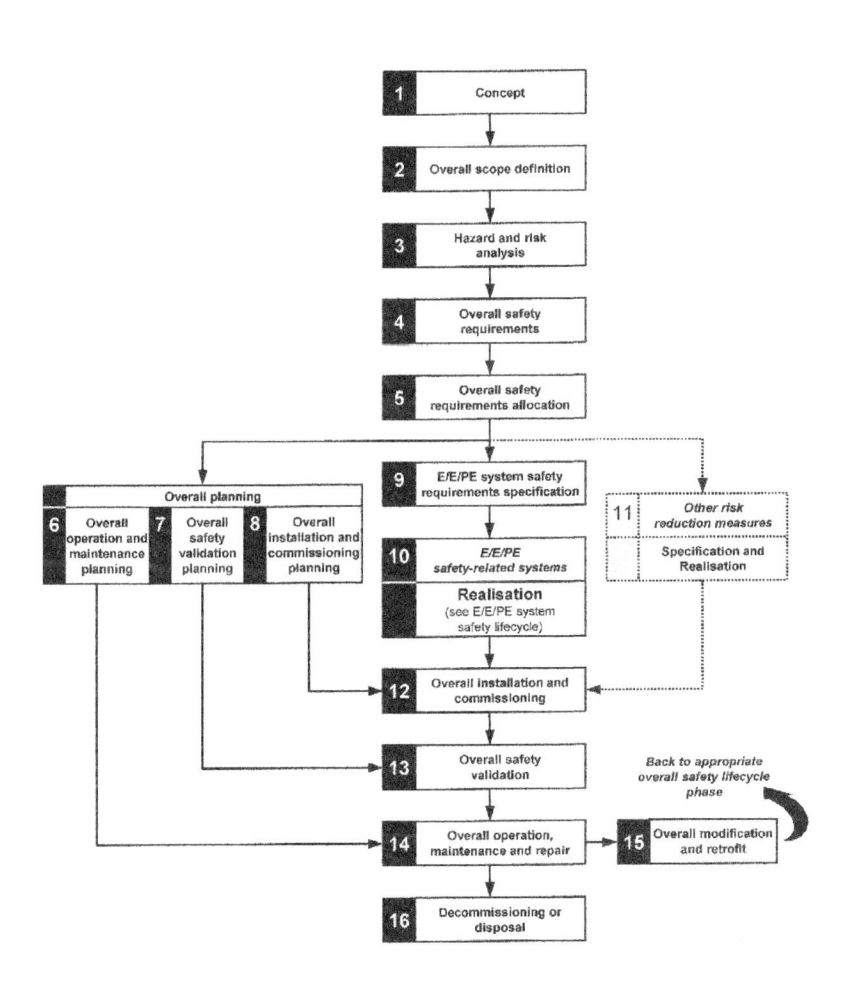

**Figure 3.1** Overall Safety Lifecycle (IEC 61508 Part 1 Fig.2)

## 4. CONTRACTUAL PROVISIONS

It is often the case that a supply contract or enquiry will stipulate as one of the conditions '...must comply with IEC 61508/61511.' This is an understandable but flawed manoeuvre that potentially makes for significant contractual difficulties. The difficulty arises because the scope of these functional safety standards is so very broad and encompasses the complete safety lifecycle associated with safety instrumented system (SIS) provisions, including (non-exhaustively); design, specification, procurement, installation, commissioning, and operation & maintenance of the system.

If procuring an engineered system (rather than a stand-alone device), compliance will necessarily require collaboration between the customer and the supplier. Many aspects of compliance may be beyond the supplier's control and much will depend on how and where the system is deployed and used. Absolute compliance is a worthy aim, but extremely difficult to achieve in practice. To be compliant in every particular, including not just the hardware and software aspects, but also all the project protocols such as planning, verification, validation, assessment, auditing etc. is a very tall order indeed (and gets taller with increasing Safety Integrity Level). Absolute compliance is approached with diminishing returns, with increasing effort required for ever smaller incremental gains in compliance. There may well be a point beyond which improved compliance would incur grossly disproportionate cost and difficulty, and not just for the supplier, since project timescales might extend significantly, and additional costs would likely propagate into other areas and would likely lead to inflation of prices.

Note that there is no legal requirement for compliance; the legal requirement (in the UK) is to reduce risk 'so far as is reasonably practicable'. That said, it should be recognised that the standards are held to constitute 'good practice' and the regulator will expect the procurement of safety instrumented system provisions to be suitably informed by the standards or otherwise to offer equivalent assurance of suitability. The real question is whether any deviations from compliance will materially affect safety and whether the overall provision is fit-for-purpose in its contribution towards reducing risk so far as is reasonably practicable.

Some may object that this introduces 'greyness' into what they would prefer to be 'black-and-white', but 'greyness' is in the nature of many engineering projects, particularly in matters of risk, and it is the role of the professional engineer to exercise responsible engineering judgement in these matters.

A key aspect of any SIS supply contract invoking the functional safety standards is the Safety Requirements Specification (this specification need not be a stand-alone document; the requirements may be distributed across a number of documents). Sometimes the requirements are not fully or explicitly identified and a supplier may be obliged to infer some requirements or assume a default consistent with his usual equipment and practices. Usually these matters are resolved straightforwardly and amicably as a project progresses, but the potential for contractual disputes is apparent.

A contract that deploys the catch-all stipulation; '...must comply with IEC 61508/61511', could give rise to disputes where a customer insists on absolute compliance despite the unwarranted cost and difficulty that this may incur for a supplier whose offering is compliant in all essentials and who has made all reasonably practicable efforts to comply. This is not to suggest bloody mindedness on the part of the customer; both parties could believe themselves to be acting in good faith e.g. the customer might mistakenly believe absolute compliance is an essential legal requirement or the supplier might not appreciate the potential significance of a deviation.

Conversely it must be acknowledged that situations might arise where a supplier might seek to avoid expense or difficulty even when they are reasonably justified, and a customer might seek compliance beyond what is reasonably practicable in an attempt to reduce or eliminate perceived liability. In recognition of these potential difficulties GAMBICA[a] offers the following model provisions for suppliers of engineered safety instrumented systems:

We undertake to comply with IEC 61508/IEC 61511 (as required), so far as is reasonably practicable, given;

- The safety requirements specified.

- Any agreement between us regarding deviations from compliance for reasons of practicability and the avoidance of grossly disproportionate cost.

- The timely availability of any necessary information from you.

- The accuracy of data provided by you.

- Any stipulations or conditions you may impose that might militate against full compliance.

- The acceptance, for any outsourced elements integrated into our offering, of suitable documented statements of compliance, where we, with due diligence, believe these to be made in good

---

[a] GAMBICA is the trade association for instrumentation, control, automation and laboratory technology in the UK. (www.gambica.org.uk)

faith by competent and reputable suppliers.

- The acceptance of suitable 'proven-in-use' (IEC 61508) or 'prior-use' (IEC 61511) arguments in demonstration of compliance for any elements that are not specifically designed and developed to be compliant with the standards.

These provisions are couched in terms of the undertaking by the supplier, but might equally be adapted as part of the stipulations by the customer. It is hoped that these terms will facilitate an appropriately collaborative venture in the procurement of safety instrumented systems. Note these model provisions are not offered as being definitive or complete; suppliers or customers should satisfy themselves of their appropriateness to any individual contract.

## 5. IDENTIFICATION OF FUNCTIONS

If we wish to manage our SIF, we must first identify them. This is not as straightforward a matter as it may first appear. I suggest that the starting point should be a cause & effect (C&E) matrix for the protection functions.

The C&E matrix will (or should) tell you what protection provisions exist on the plant/project, but typically it will not tell you what hazards are protected against; this may be self-evident in many cases, but this will not always be immediately clear.

There is also typically substantial scope for confusion. A given initiator may have the 'effect' of closing two valves. Possibly, the initiator may be protecting against more than one hazard; two independent valves may be required to close to protect against two distinct hazards, in which case the configuration would be for two functions, each 1oo1. Or it might be that closing of either valve would be successful in preventing the one hazard they were intended to protect against (e.g. two valves in series); a single function with a 1oo2 subsystem. Or possibly both valves need to close to prevent the hazard (valves in parallel lines say); a single function with a 2oo2 subsystem.

A trip may shut a valve and stop a pump; does it need to do both to stop

## IDENTIFICATION OF FUNCTIONS

the hazard, or will either do the job? Possibly either will do the job, but shutting the valve and failing to stop the pump may generate a different hazard? A critical review of the C&E matrix is needed to identify just what protection functions are implemented. You can see that this is not necessarily a straightforward matter, but the correct identification of the safety functions is absolutely key to effective management of your SIF. These can then be categorised as being for safety, environmental, or asset/business protection.

Note that sometimes a plant trip in response to an initiator e.g. pressure/temperature/level will 'do' everything; shut all valves, stop all drives, open all vents etc. but in practice many of these effects are 'nice-to haves'- they are there to 'tidy up'. In terms of supressing the potential hazard detected by the 'cause', it is usually a very restricted subset of the effects that is needed. Otherwise your final element subsystem would be 40 out of 40 or some such horribly unreliable configuration! (ALL effects being apparently required to operate to supress the hazard.)

Your critical review of the C&E matrix needs to identify the actual core safety instrumented functions (SIF); a notation scheme can be devised; F1, F2 etc. indicating individual functions together with subscripts indicating the architecture (M out of N:MooN) of initiators (causes) and final element (effects) subsystems. Note again that a given initiator-cause could be employed in more than one function and be designated e.g. F3, F6. If a 1oo2 initiator pair, there would be $2xF3_{1oo2}$, $2xF6_{1oo2}$. F3 might require a single effect to be triggered; $F3_{1oo1}$, F6 a 2oo2 effect pair; each designated $F6_{2oo2}$. The notation can be expanded to include SIL targets for each individual function.

| Cause | Effect | Close Valve 1 | Close Valve 2 | Stop Drive 1 | Stop Drive 2 |
|---|---|---|---|---|---|
| | | $F3_{1oo1}$ $F6_{2oo2}$ | $F6_{2oo2}$ | $F7_{1oo1}$ | $F5_{1oo1}$ $F8_{1oo1}$ |
| LT100H | $F5_{1oo1}$ | X | X | X | **X** |
| TT200H-A | $F3_{1oo2}$ $F6_{1oo2}$ | **X** | **X** | X | X |
| TT200H-B | $F3_{1oo2}$ $F6_{1oo2}$ | **X** | **X** | X | X |
| PT300H | $F7_{1oo1}$ $F8_{1oo1}$ | X | X | **X** | **X** |

**Table 5.1** Extract of Hypothetical C&E Matrix Illustrating SIF Designation. Note that all 'causes' trip all 'effects' but that only a subset (greyed cells) are required for hazard suppression.

Of course there is no obligation to use a C&E Matrix or to adopt a notation scheme such as suggested here. The critical thing is to identify just what needs to happen for each SIF to successfully supress the hazard; this is the core of the safety requirements specification for each function.

## 6. DEMAND MODES & DEMAND RATE

The IEC 61508 standard identifies the threshold between high/low demand functions as once per year. What is special about this figure? It represents the nominal point where there is equivalence between the PFD & PFH SIL bands where the numbers are separated by a factor 10,000 e.g. SIL1 PFD $10^{-1}$-$10^{-2}$, SIL1 PFH $10^{-5}$-$10^{-6}$. With a demand once a year (-ish, there not being quite 10,000 hours in a year) we end up with the same hazard rate with either specification.

The key point is that if the demand frequency is high relative to the proof test frequency, the testing (in a single channel system anyway) is not very useful; if demands are coming along every week there is not much point in testing every 6 months – you will know after a week that there is a problem because you will have suffered an incident. For such circumstances we have to rely on the inherent reliability of the function rather than our ability to find and fix failures in a timely manner, and so we calculate probability of failure per hour (PFH) rather than the probability of failure on demand (PFD), as the measure of the integrity of our designs with regard to random hardware failures. The limiting case for high demand is 'continuous' mode; instead of arising at intervals, a 'demand' is continuously present and the moment that a protection function fails, the hazard circumstance is realised.

That is not to say that there is no point in testing any high demand mode function; if there is redundancy in the design the function might operate successfully but be degraded because of an unrevealed failure. Testing should check for unrevealed failures in redundant channels so that they may be repaired and the system restored to full health. This consideration highlights the distinction between a 'function test' and a 'proof test'; a function test might demonstrate that the protection function is successful, but would not necessarily reveal failures in redundant channels. A properly constituted proof test should reveal failures in all channels.

Under IEC 61511, the appropriate SIL specification can be identified on the basis of either the required PFD/RRF or the required PFH. For continuous mode, we must employ PFH, since PFD has no useful meaning in that context.

For a low demand mode system, the process of SIL determination identifies the risk reduction factor required from a SIF to bring risk to a tolerable level. But if the SIF is not operating in low demand mode, RRF loses its usefulness as a performance parameter. Ultimately, in the continuous mode, we cannot ascribe a 'before' to the hazard rate and cannot therefore identify a 'before' and 'after' ratio with a corresponding RRF.

In order to determine the SIL for a high demand mode system, rather than identifying the required risk reduction, we need to identify the required failure rate (or PFH) of the SIF if the tolerable risk target is to be met. This is calculated by dividing the tolerable risk target by the combined (product) probability of failure of intervening protection layers between the SIF and the hazardous event.

Because of this difference in the way in which SIL determination is made, it is critical to first establish whether the SIF will be operating in High or Low demand mode.

If we have risk reduction layers that would normally be invoked before a demand is placed upon a SIF (typically an alarm) these may be legitimately factored into the calculation of demand rate. This is also true of any enabling condition probabilities associated with initiating events that would reduce the actual 'demands-in-anger' on the SIF i.e. those demands which would lead to the hazard consequence or demands on later layers of protection.

Note that any demands on the SIF that do not carry the hazard consequence potential are not included in the identification of the safety demand rate on the SIF. For example, it might be that some events that lead to a slow pressure rise that nevertheless triggers a high pressure trip, might not carry the hazard potential of other events that cause a rapid pressure rise that the trip is intended to protect against. There may be a useful distinction to be drawn here between 'operational' demands and safety demands. In engineering the SIF we would need to be confident that the operational demands would not compromise the ability of the SIF to respond to the safety demands; this could be a critical consideration in the preparation of the safety requirements specification. The requirement for a SIF to respond to a high frequency of operational demands might not be recognised if just the relatively low frequency of safety demands is specified.

We might then offer a refined definition of demand rate:

*The frequency of demands that, if the SIF and all subsequent layers of protection failed, would give rise to the hazard event that the SIF is intended to protect against.*

# 7. SIL DETERMINATION

There is plenty of guidance available in support of SIL determination and I confine myself here to some particular issues that are often overlooked. The mantra for all (low demand) SIL determination is this; 'what is the risk without a SIF; what does the risk need to be (what is acceptable/tolerable); how big is the gap between these two?' It is this gap, the factor by which we need to reduce the risk, which determines the SIL target for the SIF. (See Chapter 6 for details of high demand cases). If this essential is borne in mind we should not go astray. Notice that the tolerable risk is determined on the basis of the ultimate consequence i.e. if all protection layers should fail.

Identifying tolerable risk can itself be a significant challenge. The HSE identify individual risk of fatality as 'broadly acceptable' at $10^{-6}$/year. Note the qualifier 'individual'; the point is that we all individually face risks from disease, crossing the road, DIY adventures etc., and there comes a point where trying to further reduce the individual risk due to the workplace is no longer sensible. The notion is that we manage the risk in the workplace so that the hypothetically 'most exposed' individual is not exposed to an unacceptable level of risk. If the risks are sparsely distributed about a site it may well be that any given individual is only exposed to one risk at a time. On the other hand it may be that risks overlap and the individual may be exposed to different risks simultaneously. As the individual moves about a site his risk profile may be continually changing and identifying the aggregate risk may be difficult or impractical. One means of addressing this issue that is sometimes used is to reduce the nominal target by an appropriate factor to account for the simultaneous exposure possibility, but this approach is of questionable value since it suffers from the cost-benefit issue outlined below. Notice also that although analysis will typically be against a specific role e.g. plant operator, this represents an individual role, not the individual himself; in a five shift rota the individual will only be exposed for a fifth of the time. We need to be careful to distinguish between the 'risk of *a* fatality' (someone dies) and the '*individual* risk of fatality' (Fred dies); these are not the same thing and we should avoid this potential confusion when discussing or specifying risk. The shift rota dilution of individual risk may be considered as going some way towards offsetting the possibility of increased risk from multiple simultaneous hazards. (See Chapter 28 for details of risk profiling and aggregate risk calculations.)

Typically risks will be found to fall in the so called ALARP region, in which the obligation (in the UK) is to reduce the risk to be 'As Low As Reasonably Practicable'. There is no obligation to reduce the risk to the broadly acceptable level (that said, if you reasonably can, you should). The question then arises 'what is reasonably practicable'? The test is whether the expenditure required to reduce the risk further would be 'grossly disproportionate' to the benefit. In practice the ALARP condition is typically met well before the broadly acceptable threshold.

If we 'back calculate' from the broadly acceptable threshold we can see how this works:

If we postulate a risk reduction of factor 10 from an **incident** risk target of $10^{-5}$/yr, and an installation life expectancy of 25 years, with 1 shift operator at risk, the incremental gain would be $(10^{-5} - 10^{-6}) \times 25 \times 5 =$ 1.12E-3 of a life saved. If we assign a Value of Preventing a Fatality (VPF) of say £1.8M, the 'benefit' would be £1.8M x 1.12E-3 = £2025. (Note VPF is NOT the 'value of a life'; it is a notional value ascribed to reduction in risk.) If we apply a 'gross disproportion' factor of 2, we identify an investment threshold of ~£4000. It is difficult to imagine that an additional risk reduction of factor 10 (equivalent to +1 SIL) could be engineered and maintained throughout a 25 year period for such a sum.

Note that we are here now concerned with risk of *a* fatality rather than *individual* risk of fatality. That is why we here use incident frequency and number of lives at risk. In assessing individual risk we may include individual exposure time e.g. through shift rotation, of a hypothetically exposed person, but in assessing ALARP we consider the aggregate impact on all the exposed individuals.

Although a target risk figure might be lowered in recognition of multiple hazards (see Chapter 28), the additional expenditure would typically be against the individual SIFs and each would likely be found to be beyond ALARP. The lowered target would only make sense if the additional expenditure was for an aggregate reduction in risk i.e. where the expenditure 'simultaneously' reduced the risk from all the overlapping hazards. When you consider that the use of a 10-fold reduction in the tolerable risk target would turn any SIL2 requirements into SIL3, you can immediately see the likelihood of going beyond ALARP.

If we identify as tolerable an individual fatality target of $10^{-6}$/year (let alone $10^{-7}$/year!) we will be pretty much doomed to having to make an ALARP demonstration for every SIF. It may be more sensible to use a less onerous initial target risk target for SIL determination purposes and then ask whether anything further could be done without grossly disproportionate expense. If we used a target of $0.2 \times 10^{-5}$/year, the notional investment threshold (using a gross disproportion factor of 3) against a single fatality incident might be £1200 as identified above. If we used $0.2 \times 10^{-4}$/year it might be £12,000. You might then ask what is the point of the $10^{-6}$/yr threshold. The point is that if you find the risk to be already at or below this target, then you do not need to consider whether there is anything further that might be done to reduce the risk. Above this threshold you must.

In the early days of IEC 61508, the Risk Graph approach was very popular, but anecdotally anyway, it appears less so now and there is wider take up of Layers of Protection Analysis (LOPA) as an alternative. One issue with the Risk Graph is that the model shown in the standard is not particularly helpful; it is a limited example only of the approach and needs developing to provide a satisfactory tool for use. The number of branches and the number of levels can be adopted to suit the particular requirements of the user. The graph will also need 'calibrating' i.e. it will need adjusting to match the SIL target 'outputs' with the tolerable risk targets for the enterprise being assessed. This is critical to satisfactory use of the approach and must not be overlooked. It is helpful to then take the graph for a 'trial run' and evaluate the findings for some representative scenarios. It is also useful to develop some guidance that can help steer the users through the assessment by providing criteria against which questions such as 'frequency of exposure' or 'probability of avoidance' may be evaluated. If properly developed and calibrated, the risk graph is a perfectly acceptable approach, and has the advantages of being relatively simple and quick to use, and of being readily understood by those such as operators who may not be familiar with SIL concerns. The disadvantage is that it will always offer limited resolution in the discrimination of risk. Attempts have been made to overcome this by introducing more sophisticated versions of the risk graph, but these start to look increasingly like a LOPA. There may well be a tendency towards so called SIL 'inflation' if the user is prompted to make a succession of conservative evaluations at each branch point 'to be on the safe side'; this compounded conservatism can quickly drive SIL targets to spuriously high levels. This can lead to an unwarranted degree of pain and grief in the engineering. For these reasons, risk graphs are increasingly used as a 'first pass' filter in SIL determination; if the result is SIL1 (possibly 2 if evidently appropriate) or less the result is accepted, but anything higher is subjected to a more rigorous evaluation approach. The same points may be made in relation to the risk matrix approach.

If using a qualitative approach, strictly the PFD/PFH target should be nominated to correspond with the low end of the SIL band (see Chapter 1).

LOPA is a widely misused tool; the principle is straightforward enough, but there are some subtleties in its use that are often overlooked:

- If we postulate failure of the Basic Process Control System (BPCS) (typically a control loop failure leading to a process excursion) as an initiating event, we cannot then legitimately claim the same control system (or an alarm derived from it) as a defence against such an excursion. This is so called 'double dipping'; effectively claiming twice for the same provision.

- If the scenario proves to be 'high demand' the usual LOPA (low demand) risk reduction model is no longer valid; the calculation should be of the required PFH, NOT the required RRF. (Note that many LOPA tools do not make this distinction.)

- If the potentially hazardous event is such that operators would be drawn to the hazard zone (e.g. to investigate an alarm) then claims for low occupancy may be invalidated; effectively there would always be an operator present when the hazard potential is present.

- Any claim for operator intervention suppressing a hazard in response to an alarm will only be valid if there is a clear means by which the operator may intervene effectively and there is a sufficient time between the prompt and the actual hazard for him to act. If the hazard would be realised within 1 minute of the alarm, for example, then it would not be considered realistic to claim a reliable response from the operator.

- So called enabling events or conditional modifiers are sometimes overlooked, but these may allow legitimate claims for risk reduction. A typical example is the weather conditions prevailing at the time of a flammable release; it is only in the absence of a good stiff breeze that any explosive atmosphere hazard potential would arise. If it can be demonstrated that relatively still conditions only arise 20% of the time, then a factor 5 risk

## SIL DETERMINATION

reduction against explosion hazard would be entirely legitimate. Conditional modifiers are typically assigned as mitigation layers that prevent the hazard event realising the hazard consequence e.g. a release that does not lead to an explosion. These are typically 'universal' provisions that apply to all initiators e.g. weather or occupancy. Failure to discriminate between these circumstances may distort the analysis. (See also Chapter 41: Consequence Mitigation in LOPA.)

- In the absence of 'enabling conditions' a demand on the SIF would not arise. If a plant only operates a hazardous process intermittently, say 50% of the time, an enabling condition probability of 0.5 may be assigned. These conditions may well relate to some initiating events but not others.

- Zoning of areas, with control of ignition potential through 'Ex' protection mechanisms, will allow a risk reduction claim against minor flammable material leaks, but may not be useful against catastrophic releases. Note also that if the initiating event will itself present a source of ignition e.g. machine disintegration, then no claim may be made.

- If a release is both flammable AND acutely toxic then the control of ignition sources could not be claimed to reduce the risk to unprotected operators.

- Claims for the likes of mechanical relief valves or non-return valves will only be legitimate if their inspection and maintenance provisions can be seen to support the claim; these items cannot be considered as 'fit-and-forget'.

- Claims for risk reduction through operator intervention (typically in response to an alarm) are usually for no more than a factor 10. Exceptionally, if it can be shown that all the ergonomic considerations substantiate a greater claim, it may be possible to claim up to a factor 100.

- A claim for a storage tank bund might be useful against an overfill cascade from a storage tank in respect of environmental hazards, but not for the hazard arising from formation of a vapour cloud.

In calculating initiating event frequency it may be that improved human error probabilities may be claimed; there is a difference between operator reliability 'on demand' when there is no discretion in the timing of the execution, and operator performance 'on election' when he is executing a well prepared, established procedure, with timing at his discretion and the option of halting the procedure if he so chooses. (See Chapter 39: Human Factors.)

Fault Tree Analysis is a more sophisticated and rigorous tool sometimes employed for SIL determination; it is correspondingly burdensome and is not really suitable for routine use. It is typically reserved for the more hazardous and complicated situations where high SIL functions might be required. That said, for those with the appropriate understanding, it is not so very difficult to undertake for smaller scale studies and it may be a useful possibility.

## 8. SIMPIFIED LOPA

As an intermediate approach between a simple risk matrix/graph and 'full blown' LOPA, a simplified LOPA or what I call 'SLOPA' might be adopted for low demand mode functions. The idea is that we consider each SIF in turn and consider (and record) first of all what initiating events might place a demand upon the function. A judgement is made about which initiating event interval would be most appropriate:

0.3 – 3.0 years

3.0 – 30 years

30 – 300 years

300 – 3000 years

3000+ years

Notice with this banding, 1 year, 10 years etc. sit in the geometric centre of the bands. This is particularly helpful when dealing with the likes of control system failures which are typically estimated at 0.1/year; an interval of 10 years.

We then consider what conditional modifiers (e.g. weather/occupancy) and other defences (e.g. mechanical relief) could be legitimately claimed against the initiating causes. For each modifier/defence a judgement is made about the order magnitude risk reduction that may be claimed (10, 100, etc.).

Multiplying the demand interval by the identified risk reduction factors gives us the corresponding SIF demand 'in anger' interval which can be entered using the banding as before. This is the period with which we anticipate a demand on our SIF, where, if it fails, the ultimate consequence will be realised. Let's call this a 'critical' demand. (Note this is not the same as how often the SIF will be invoked.)

The ultimate consequence is then categorised, typically; fatality (I), serious injury (II), restricted work (III), first aid (IV), no hazard (V). For each of these categories a previously identified tolerable interval is allocated. A look up table can be created to allocate the SIL target on the basis of the consequence category and the SIF demand interval. This table establishes the calibration of the tool.

| SIF Critical Demand Intervals | Consequence Categories ||||| 
|---|---|---|---|---|---|
| | I | II | III | IV | V |
| 0.3-3.0 | 3 | 2 | 1 | 0 | - |
| 3.0-30 | 2 | 1 | 0 | - | - |
| 30-300 | 1 | 0 | - | - | - |
| 300-3000 | 0 | - | - | - | - |
| 3000+ | - | - | - | - | - |

**Table 8.1** Example SIL target Allocation Look-Up Table

If we identify a consequence category I hazard with a nominal 100 year SIF demand interval, a SIL1 target would be allocated. With a SIL1 risk reduction of factor 10-100, the hazard interval would be 1000 -10,000 years. If individual exposure was 20% of the time, (continuously exposed when on shift) then the individual risk would be in the 5,000-50,000 years range (nominally 17,380 years at the geometric centre). (All figures for purposes of illustration only and should not be regarded as definitive.) If wanted, the same approach can be used with environmental and business hazards. In which case it is necessary to identify which of the modifiers would apply to these hazards e.g. occupancy would not apply as a modifier for an environmental hazard. A simplified illustration which shows the essentials of the approach is shown overleaf.

| SIF | Initiating Event | Initiating Event Interval | Other Defences | SIF Critical Demand Interval | Consequence Category | SIL | Notes |
|---|---|---|---|---|---|---|---|
| High Level Trip | Control Failure | 3.0-30 | Occupancy (RRF 10) Zoned area (RRF 10) | 300-3000 | II | 1 | Potential fatality if operator present (10% occupancy). Release would be to zone with control of ignition sources (0.1 ignition probability). |

Table 8.2 Example of SLOPA Line Entry

As discussed in the previous chapter on SIL Determination, SLOPA might be usefully deployed as a first pass filter. If it spits out SIL2 or above it may well be appropriate to employ a more rigorous analysis tool to verify the result.

## 9. EVALUATING SIF ACTION FAILURES

When undertaking a HazOp studies it is understood that in examining the implications of a deviation in a given node, we don't postulate completely independent simultaneous failures elsewhere in the process; sometimes referred to as 'double jeopardy'. There is a very good reason for this; the possibilities expand fantastically (and it is not as though HazOp is quick, cheap and fascinating!)   But there are circumstances where we **should** postulate equipment failure in other nodes at the same time as our primary deviation occurs – SIF operation.

The point is this; when a deviation occurs we may invoke a SIF to protect against the potential hazard.  The SIF may trigger a multiplicity of 'effects' in response to the deviation 'cause' and we will be at pains to engineer this SIF with the appropriate integrity in accordance with the SIL target that has been identified.  Now it may be that one of those 'effects' fails and gives rise to a new hazard that is quite separate from the initiating deviation.

A simple example would be a high level trip of a pump that stops the drive and closes the discharge valve.  If the valve closes, but the drive fails to stop, we will be faced with a new hazard; the potential over-heating of the dead-headed pump.  This might require the engineering of separate protection/interlock functions or other defences. (This particular example might well be caught by the conventional deviation 'flow – none', but you can see the potential for impacts in remote nodes.)

Another example is a compressor that trips on low seal oil level with closure of suction and discharge valves and opening of the vent valve. If successful this will prevent a potential gas leak from loss of sealing. However, if the suction valve should fail to close, a new potential hazard arises with the venting of the upstream plant inventory, rather than just the machine.

So the question arises, how are these possibilities to be evaluated? If a C&E matrix (or equivalent) has been generated we can consider the operation of each SIF and postulate the failure of individual 'effects' when the SIF is triggered. In our pump example, we have already examined the possibility of successful valve closure with a failure to stop the drive. The other possibility is successful stopping of the drive and failure to close the discharge valve, leading to possible reverse flow scenarios.

If there are 10 effects, we need to examine each in turn to identify the possible consequences of failure when the other 9 are successful. Very often, no hazard will be identified and most effect failures can typically be quickly eliminated as of no particular consequence, and so the exercise is not inordinately burdensome. Note however the requirement to consider all the effects, not just those that are required to supress the hazard associated with the cause. Even the failure of effects that are not part of the SIF proper may give rise to a new hazard.

The consequence should be assessed and categorised as usual and the demand rate/interval is identified from the frequency with which the SIF is expected to be triggered. Note that the concern here is with the actual trigger frequency, rather than the frequency with which the SIF is called upon 'in anger' (where if the SIF fails to respond the ultimate consequence would be realised). The risk reduction requirement will then be identified in the usual way from any gap between tolerable frequency of the consequence and the frequency with which the SIF will be triggered. (Note that 'SLOPA' (Chapter 8) may well be a suitable tool for this evaluation.) This will identify the SIL requirement on the individual effects, which strangely, may even be higher than the original 'core' SIF. Many of these secondary SIFs, which consist of final element subsystems only, may prove to be 'high demand' in that they may be called upon more than once per year.

The same concern does not arise with spurious operation of individual effects since these should be addressed as conventional deviations (single jeopardy?) in the relevant nodes during HazOp, for example, spurious closure of a valve should be captured under flow-less/none.

## 10. RISK ASSESSMENT TOOLS

Risk assessment tools should all come with a health warning; 'uncritical use may seriously damage your business'. Worst case assumptions may appear prudent, but their undiscriminating use may seriously undermine the value of a risk assessment in identifying the appropriate allocation of resources. There are a variety of approaches that may be adopted in undertaking risk assessments in support of the development of a safety case or Safety Integrity Level (SIL) determination. There is a spectrum of possibilities ranging from the purely qualitative, through semi-quantitative to the fully quantified. In assessing risk the HSE employ the concept of 'proportionality'; risk assessments and adopted risk reduction measures are required to be 'proportionate' to the risk. In essence, the greater the risk;

a) the greater the required degree of rigour in the risk assessment and the demonstration of ALARP and;
b) the more a duty holder would be expected to pay to reduce those risks.

A risk may be characterised as having 'high proportionality' if it is assessed as approaching the intolerable region of the HSE's 'risk triangle', which categorises risk as being 'broadly acceptable', 'tolerable if ALARP' (As Low As Reasonably Practicable), and 'intolerable'.

Proportionality should not be confused with 'proportion factor' which is the ratio of the Cost of Preventing a Fatality (CPF) and the Value of Preventing a Fatality (VPF) and which may be used in assessing whether any further expenditure on incremental risk reduction would be 'grossly disproportionate' as a test of whether a risk was ALARP.

Some of the available risk assessment techniques are highly refined, but this should not blind us to the inevitable uncertainties in the results. A rigorous process will not compensate for uncertain data and any quest for absolute accuracy is doomed. A particular problem is the compounding of conservatism: A succession of 'worst case' assumptions that are thought to represent an upper bound to risk will quickly generate gross distortion of the overall assessment. It may well seem prudent to 'err on the side of caution', but do this at successive points and your analysis may well be one or more orders of magnitude from a 'true' estimate. It might be argued that the distortion is 'safe' but if it causes a misdirection of resources or an unwarranted elaboration of provisions which are more difficult to manage/maintain, the result may well be an overall net loss of safety in practice.

A wrongheaded and naïve approach is to perform a risk assessment and unquestioningly accept the outcome as definitive. The outcome should always be critically reviewed to see if it appears **sensible**. It may be that the assessment reveals hitherto unrecognised levels of risk or identifies an element as being of more significance than was previously understood. It is perhaps when revealing any such anomalies that these tools are at their most useful. But if the outcome does not fit with your experience or seems to suggest that what you understood to be established good practice is inadequate, you should probe further to discover whether it is your expectations or the assessment that is flawed. If your SIL determination process reveals wholesale deficiencies, your first thought should be to check the calibration of your approach.

The tools at the less rigorous end of the spectrum e.g. risk graphs and matrices, tend to be relatively coarse in their resolution of risk and are more susceptible to inadvertent compounding of conservatism. These less rigorous tools might be effectively deployed as a first pass filter with apparently higher risk scenarios 'parked' for a higher resolution and higher rigour assessment. The key point is that risk assessment should be performed to validate your judgement (which should itself be suitably informed by training, experience, knowledge of good practice and an understanding of pertinent mechanisms) rather than to attempt to directly and definitively establish the 'true' level of risk.

It in recognition of this that the 'Guidance on Risk for the Engineering Profession' published by the Engineering Council [1], says:

- *bear in mind that risk assessment should be used as an aid to professional judgement and not as a substitute for it*

It may be that after critical probing to find the reason for any discrepancy between your expectation and the assessment, you may be obliged to 'recalibrate' your judgement; embrace the opportunity, this is quality CPD. The 'new' risks must then be addressed in an appropriate, responsible manner. Clearly what is not acceptable is to cynically manipulate the risk assessment to remove any embarrassment or inconvenience.

On the other hand, your critical review may find that the assessment is flawed in some way; perhaps through inadvertent compounding of conservatism or some critical omission. Perhaps a missing conditional modifier or enabling event; some circumstance that should be factored in to the analysis to qualify the true hazard potential. E.g. a flammable release risk should be qualified on the basis of; the likely size of the release, the presence and vulnerability of people, the likelihood of a source of ignition, the likelihood of detection and effective intervention, the likelihood of the absence of a good stiff breeze etc.

Very often it will be self-evident that the probability of a given circumstance will be less than 100%, but there may be no ready means of evaluating how much less. The default position is therefore to conservatively assume a probability of 100% – not unreasonable in itself, but if repeated a number of times we may suffer gross distortion in our final estimate. It may be that in recognition of this 'default' conservatism we might choose **not** to be conservative in respect of some other provision which is more tractable in estimation.

A good example is the guidance given in the Buncefield report (ref 1) in terms of the 'Probability of explosion after ignition' as a conditional modifier to be used in LOPA studies (Appendix 2, clause 141):

*'Given the present state of knowledge about the Buncefield explosion mechanism this report tentatively proposes that the value of this modifier should be taken as unity in the stable, low wind speed, conditions that are the basis of this hazardous scenario. A much lower, and possibly zero, probability might be appropriate. It is possible that an improved understanding of the explosion mechanism may allow a better basis for determining the value of this factor in the future.'*

The explosive (as distinct from flash fire) nature of the Buncefield incident was something of a surprise; the implication would seem to be that the circumstances were unlikely otherwise there would have been no surprise; the phenomenon would have been well known and the possibility recognised.

If the actual probability does indeed approach zero, the assessment of the risk associated with an explosion will have been inflated by a factor 'approaching' infinity!

This highlights the need to be wary of accepting risk assessment quantification as providing any measure of absolute risk; these tools are typically more useful in providing a measure of relative risk and providing a means of comparison. This is of particular significance if you are to perform any cost benefit analysis, since this requires that you postulate 'before' and 'after' levels of risk and the corresponding cost. Expenditure typically secures an additional risk reduction factor (rather than an absolute amount of reduction) and the apparent benefit will therefore vary with the identified level of risk 'before': There is more benefit from a factor ten reduction in the annual probability of a hazardous event from $10^{-2}$ to $10^{-3}$ than there is from $10^{-3}$ to $10^{-4}$.

It is all very well taking the anti-Panglossian[a] view of 'all for the worst in the worst of all possible worlds', but we need to be smarter than that when deciding how to allocate limited resources. If there are a number of provisions in the analysis that are known to be conservative by indefinite amounts, it may be appropriate to aggregate these and assign a factor to their combined influence in order to counter the distortion that would otherwise arise with compounding conservatism.

You may find that an iterative development of both your understanding and the assessment is required until your increasingly educated expectations converge with the increasingly refined analysis. Far from being a 'massaging' of the numbers, this is the proper and responsible approach of the professional engineer. There is a world of difference between modifying a risk assessment because it is not judged sensible and because it does not fit the budget.

If it comes to a debate with the HSE about a risk assessment or safety case, remember that their inspectors will approach these questions from a distinct perspective. Understandably, this is unlikely to embrace your wider business concerns and resource constraints. It is for the duty holder to make a robust case that also recognises these issues. The HSE may well urge you to conservatism, which undoubtedly has its place in a prudent and responsible approach, but if you find this to be leading to unwarranted distortion you should stand ready to defend your judgement.

**References**
1. Safety and environmental standards for fuel storage sites. Final report. HSE 2009

---

[a] Pangloss being the character in Voltaire's *Candide* who continually asserts that 'all is for the best in the best of all possible worlds

## 11. COMPETENCE

With the introduction of the 2$^{nd}$ edition of the functional safety standard IEC 61508, the requirements for competency became mandatory i.e. a necessary condition for compliance. The questions: 'What are the requirements for competence?', 'On what basis is competence to be assessed?' have taken on new force. A range of training courses are promoted by a wide range of suppliers, but the (possibly uncomfortable) truth is that there is much more to competency than attending/passing a short course.

The IET has published a code of practice ('in collaboration with the HSE') for 'Competence for Safety-Related Systems Practitioners' [1]. This is a reworking of earlier guidance developed relatively early in the history of the 61508 standard following collaboration between the HSE, the IET and the British Computer Society. The principles the guidance expounds are sound enough, but many have found the specific model offered is too ambitious and complicated for practicable implementation.

The guidance proposes the identification of 'roles' with required competence profiles, for which an identified level of competence is required in respect of each of a range of associated activities and personal attributes. Although the identification of roles may appear to be a natural approach (in that they may be mapped by the project organogram), it may be found that it makes for an inflexible system, with unduly burdensome administration, that struggles to manage the evolving requirements of a project and its staffing. The key requirement is, of course, to be satisfied that each activity is performed competently; the designated role of the performer is less of a concern.

**Competence Levels**

The guidance identifies competence levels as:

Level 1 (formerly 'Supervised Practitioner'): *"Someone with Level 1 competence in a defined task-based or supporting competency has sufficient knowledge and understanding of good practice to be able to apply those competencies without placing an excessive burden on the more competent safety practitioner who is responsible for supervising their work. Level 1 competence may have been developed through targeted training and work on non-safety-related projects."*

Level 2 (formerly 'Practitioner'): *"Someone with Level 2 competence in a defined task-based or supporting competency has sufficient knowledge and understanding of good practice, and has the demonstrated experience, to be able to apply those competencies without the need for supervision. They will maintain their knowledge and be aware of the current developments in the context in which they are applicable."*

Level 3 (formerly 'Expert'): *"Someone with Level 3 competence in a defined task-based or supporting competency has sufficient understanding of why things are done in certain ways and has the (sic) sufficient managerial skills to be able to take overall responsibility for the performance of a task or activity. They will be familiar with the ways in which systems, and previous safety management systems, have failed in the past. They will keep abreast of technologies, architectures, application solutions, standards, and regulatory requirements, particularly in rapidly evolving fields such as programmable safety-related systems. They will also have sufficient breadth of experience, knowledge and understanding to be able to work in novel situations."*

We might add a further Level 0 ('Novice') for those that have little or no established capability and have yet to progress to Level 1.

For ease of use, I retain the former competence designations in the remainder of this discussion. (Their abandonment strikes me as perverse).

In the context of competence management, the designation 'Expert' may be of limited value: by definition, anyone assessed as a 'Practitioner' for a given task/activity must be capable of executing that task competently. It is not clear that 'super competence' would add anything in that respect. If the task is a broader activity such as 'management', then the criteria for 'Practitioner' level in that task should have corresponding breadth. The designation may be useful for other purposes; e.g. to indicate someone with a broader range of competencies or the ability to exercise a given competence across a wider range of contexts, but this is beyond the immediate question of task specific competence.

The guidance rightly points out that what constitutes competence will depend not only on the task or activity, but also the 'context' in which the task is to be performed. Competence in risk analysis in the aerospace sector say, would not necessarily translate to corresponding competence in the process sector. Competence with relay technology would not necessarily translate into competence with a PLC system.

Rather than attempt a high degree of 'granularity' in the specification of competence criteria and assessment, a simpler approach is to employ assessment on a higher level generic basis, looking for underpinning knowledge and understanding, together with experience in an appropriate context, and then, crucially, to couple this with specific work execution protocols (as outlined in the next section). This assessment to be on the basis of;

- Relevant professional/trade qualifications
- Relevant training (training may be useful, but will not on its own provide assurance of competence; I have seen 'certified' practitioners I would not pay with buttons! Consider: everyone with a driving licence is a 'certified' driver!)
- Relevant experience (positions held, work undertaken)

- Currency of experience and degree of professional engagement
- Personal attributes

It is difficult to assess personal attributes, such as diligence and professionalism, other than by direct experience. References may be useful. Registration as a professional engineer requires demonstration of appropriate communication and inter-personal skills and, in the absence of contrary evidence, might allow a presumption of appropriate attributes.

The key point is not to presume competence or to accept claims without evidence, but to establish an auditable record of the basis on which competence is assessed. This might include a CV with a specific focus on functional safety (a 'statement of functional safety competence') together with related CPD records and supporting evidence, possibly including interview reports or testimony from references. These records might then provide an ongoing basis for periodic competency appraisal and identification of training or other development requirements; this might be a subset of a wider performance/competency appraisal process that is not confined to matters of functional safety.

**Work Execution Protocols**
Managing the competency of individuals is a means of gaining assurance that the work they perform will be competently executed; ultimately it is the work that matters, not the individual. Individual competence is certainly important, but competent individuals can still make mistakes or be misdirected. As a complement to the assessment process described above, the following organisational and project protocols can be adopted/adapted to provide a wider basis for the management of competent work execution:

- An initial default assumption (unless the evidence is unequivocal), that individuals are competent to no more than Supervised Practitioner level for any given task

- All work to be subject to formal check and approval by persons assessed as Practitioners (or Experts)

- Supervised Practitioners only to be promoted to Practitioner for a given task when they have had sufficient of their work assessed as satisfactory, typically through the check and approval protocols. (And that this designation be revoked if contrary indications arise.)

- Check and approval protocols to be explicitly identified, detailing just what is expected from checkers and approvers. These protocols to require that work submitted by a Supervised Practitioner (or novice) must be subject to suitably comprehensive and probing review.

- It is made clear that individuals are accountable for their work; sign off provisions to be implemented with appropriate discipline. (Accountability is a powerful means of promoting diligence.)

- Authority for check and approval to be conferred on the basis of competence assessment as identified on a competence register. It might be useful to compile a project specific matrix identifying activities (e.g. compile Safety Requirements Specification, perform Layers of Protection Analysis Study, use a particular Probability of Failure on Demand calculation tool) against personnel, with prevailing competency level entered for each intersection. A third dimension of 'context' may be added if useful.

- An undertaking from individuals that they will not act beyond the limits of their competence (no guessing allowed). This is a key attribute. It is of course a requirement of the codes of conduct of the professional engineering institutions, but it is perhaps useful to ask for an explicit undertaking in respect of functional safety.

- The employment of document templates and other project tools that provide appropriate structures, guidance and prompts to promote appropriate coverage and effectiveness.

- Independent functional safety assessments i.e. not by those performing the work.

- Audits of protocol execution/adherence.

## Conclusion

It is hoped that the outline given above will provide a relatively simple and practicable basis for the management of competence in matters of functional safety, or indeed for other areas with comparable requirements.

## References

1. Code of Practice Competence for Safety-Related Systems Practitioners, IET, 2016

## 12. OPERATION & MAINTENANCE

Functional safety practices, as promulgated by IEC 61508, have now been with us for many years, but much of the focus has been on the Safety Integrity Level (SIL) determination and realisation phases (design & installation). These are critical aspects of course, but it is interesting to reflect that a typical safety instrumented function will spend the overwhelming proportion of its life in the operating and maintenance phase of the safety lifecycle; this is where the failures really matter, but it does not perhaps receive the attention that it deserves. A perfectly sound installation is unlikely to remain so if appropriate maintenance practices are not adopted. So what are these practices and the related training requirements? We may identify them as belonging to the following distinct areas:

- Proof testing
- Inspection
- Repair
- Modification
- Security
- Performance Monitoring

## Proof Testing

Proof test procedures must be suitably rigorous and comprehensive. They should be fully documented and perfectly explicit, with appropriate pass/fail criteria specified: highly generic 'procedures' such as "Calibrate and test trip function PZH123" will not do. Those performing the tests should understand their responsibilities and their accountability for proper execution. They should understand that a procedure may stipulate that a test be done in a particular way to achieve the proper test coverage and that 'short cuts' should not be used unless approved by a responsible and competent authority. The test record should distinguish between 'passed first time' and 'repaired and then passed'. (See Chapter 25 for further details.)

## Inspection

Technicians should understand that there is a separate requirement to inspect Safety Instrumented Functions (SIF) for signs of degradation; a SIF may function perfectly, but if it is showing signs of distress this must be reported to the maintenance authority to allow any implications for the wider population to be assessed and for rectification work to be put in hand.

## Repair

It should be understood that any repair, other than like-for-like exchange, should be approved by a responsible and competent authority. After any repair, a test must be made that covers all aspects of the SIF that could be impacted by the repair. A full proof test may well be appropriate.

## Modification

Any modification to a SIF must be subject to appropriate change control procedures. These must ensure that any modification does not compromise the function's ability to meet the stipulations of the safety requirements specification. Emergency 'frigs' (*ad hoc* overrides) are not acceptable; any override must be assessed and authorised by a responsible and competent authority. (See Chapter 31 for further details.)

## Security

As far as is practicable, SIFs should be secured from unauthorised interference. Access should be controlled e.g. password protection employed and system cabinets locked. Technicians should understand that security provisions should not be bypassed or disabled and that unauthorised personnel should not be given access. It is advisable that all SIS devices, cabinets and junction boxes etc. are clearly identified as being SIF equipment and it should be understood that no unauthorised access, adjustment or modification is permitted. (See also Chapter 34).

## Performance Monitoring

Failure of any SIF should be fully investigated to identify the reason for the failure. Failures should be reported to the responsible maintenance authority through an established reporting procedure. This will allow assessment of whether there may be implications for other functions. Technicians should be briefed to recognise and support the need to monitor the ongoing performance of SIFs to validate design assumptions and to identify possible equipment end-of-useful life issues. (See Chapter 29 for further details of performance monitoring and FSA4)

## Training Requirements

Any instrument craftsman deserving of the name will understand how to install, maintain and repair instrument equipment and systems. The training requirement in respect of the operation and maintenance of SIFs revolves around an awareness of the significance of SILs and an understanding of the aspects outlined above. For the operation and maintenance phase, it is not necessary to provide training on the niceties of SIL determination, SIF design and associated quality assurance and management provisions: given an underpinning knowledge of instrumentation systems, the training requirement is relatively limited and could typically be adequately addressed in ½ -1 day. It is essentially about making sure technicians understand the need for appropriate discipline in the maintenance and operation of SIFs and that a 'make-do-and-mend' approach is not acceptable. If this foundation of awareness and understanding can be established, the need for extended courses and refresher programmes may be questionable (the same may be said of many areas besides functional safety; many courses are inflated beyond the users' actual requirements). Indeed, if periodic refresher programmes were found to be necessary to promote proper conduct, it might indicate a more fundamental concern with personnel competency; a want of inherent capability or discipline. (We must distinguish here between 'refresher' and 'update' training.)

**Conclusion**

In summary then, for training of technicians for the operating & maintenance phase of the safety lifecycle, the training course outline should include:

- The philosophy and scope of the functional safety standards
- The significance of SILs
- An overview of the Safety Lifecycle
- Proof Testing philosophy and approaches
- Maintenance/repair/modification considerations
- Performance monitoring and fault reporting
- Competency, personal responsibilities and accountability

And, if the course is in-house;

- Site specific issues/practices

In-house training must address, by way of risk assessed method statements, any safety procedures and safeguards that must be adhered to when taking live plant to its trip condition in order to carry out a proof test.

## 13. CERTIFICATION & PRIOR-USE

Let's dismiss one common misconception immediately: the standards do NOT require certification of anything: not equipment, not people, not companies. Certification is one way of acquiring confidence in fitness-for-purpose, but it isn't the only way and it is not necessarily the best.

'Prior-use' is a term used within the IEC 61511 standard for functional safety systems in the process industry sector, and refers to the determination that a component or subsystem is suitable for deployment in a safety system on the basis of established experience, rather than on the basis that the equipment is '...in accordance with IEC 61508-2 and IEC 61508-3, as appropriate...'.

What this boils down to for most end users, is the selection of equipment based on established experience rather than certification by others. As you might expect, this is of particular relevance to legacy plant installations.

IEC 61508 uses the corresponding term 'proven-in-use', but the requirements to demonstrate 'proven-in-use' (PIU) in accordance with IEC 61508 are different from those required to demonstrate 'prior-use' in accordance with IEC 61511. (If I may paraphrase, PIU requires 'clearly restricted functionality' and 'adequate documentary evidence' based on sufficient operational experience in conditions that are 'sufficiently close' to those for the proposed SIS deployment.) Be wary of those that attempt to 'pick and mix' these requirements. This is horribly confusing territory. The terms are synonymous in terms of concept, but **not** in terms of requirements.

It may be acceptable to point to a reputable vendor's claim of proven-in-use (under IEC 61508) as substantiating a claim of 'prior-use' under IEC 61511, but you would need to be satisfied that your equipment deployment was not significantly different from that of the population used to establish 'proven-in-use'. An onerous duty is not likely to be adequately represented in a wider population study; it is likely that a benign duty would be.

There are any number of people that will queue up to tell you how hard the prior-use (PU), argument is to deploy. (Most with an interest in certification.) Some shameless scaremongering has been employed. Is it really so difficult? Consider; why would the standard committee include the PU provision and then make it next to impossible for the user to adopt as a practicable option? The compilation of PU substantiation is not necessarily that onerous, particularly for SIL1 and 2.

Some people seem to regard prior-use as the poor relation of design certification, but this is not correct; good experiential evidence is to be preferred to theoretical prediction and is ultimately the only means of validating reliability modelling. You may also find that suitably certified equipment is simply not available; your only recourse will be to a prior-use argument. Certification is itself of very variable quality. Some certificates only address the hardware and do not cover any software deployed; which inevitably makes the certificate of limited value to the user. Some certificates may use unrealistic operating/installation assumptions. Even when certified, a critical review may reveal deficiencies in the certification process.

The PU provision is included in the standard in recognition of the practical considerations and constraints in the 'real world'. The key requirement is to establish with an appropriate degree of confidence that the equipment selected is fit for purpose.

In this respect, IEC 61511-1: says:

The evidence of suitability shall include the following:

- consideration of the manufacturer's quality, management and configuration management systems;

- adequate identification and specification of the components and subsystems;

- demonstration of the performance of the components and subsystems in similar operating profiles and physical environments;

- the volume of the operating experience

So how may we proceed? The first couple of items are relatively straightforward and should pose no particular difficulty with any reputable manufacturer; a traceable record of the pertinent findings and specifications needs to be made. This is likely to include any accreditation such as ISO9001 or TickIT, or identified policies or procedures. Also complete identification of model and version numbers, including where appropriate (i.e. where judged to bear on performance) any manufacturer build version numbers (hardware and firmware) as distinct from what we might call the catalogue reference (used for procurement). The remaining two items are less straightforward, but remain practicable.

Equipment that is not process wetted and that typically sits in marshalling/auxiliary rooms (e.g. contactors, trip amplifiers, relays), will typically be less sensitive to duty than would be the case for process wetted items. It may be correspondingly relatively straightforward to establish whether the equipment is suitably rated for the environment. Failure data for a significant equipment population may be available from the vendor or from within your business or within industry databases. Vendor data based on equipment returns will likely be optimistic, since it does not include for equipment that ends up in the bin. Returns made within the warranty period may offer a more realistic basis, but are still likely to be optimistic – It may therefore be appropriate to apply a nominal factor to account for this possibility, one authority (Dr D. J. Smith, author of the FARADIP database, reports that he finds it prudent to use a factor 5 to correct for over optimistic reporting.) Note that the data needs to be for the model and build that you propose to deploy, or you need to be suitably confident that any difference will not have compromised the data. This is a potential difficulty because you may not have access to the necessary details of the software revision history (and not all vendors will keep appropriate records). For the most part, the practicable approach will be to satisfy yourself that the manufacturer is reputable and has adopted the appropriate procedures for managing these concerns, which brings us back to the first part of our evidence portfolio.

Remember also that your system reliability calculations should also include for any unreliability associated with the process connections themselves; it is all very well having a failure rate for a pressure transmitter (say), but if this has been determined from a Failure Modes and Effects Analysis (FMEA) of the electronic assembly, it will not include for failures of the physical system associated with the impulse line connection for example. (See Chapter 20) This is one reason why good data from operating experience is to be preferred.

For process wetted equipment, the failure rate is likely to be susceptible to the duty. If the duty is relatively benign e.g. no environmental extremes, non-fouling, non-corrosive/erosive etc. then failure rate data from a wider population that does not distinguish between duties (and therefore includes more hostile duties) will be correspondingly conservative and might be used without qualification. If the duty is hostile, judgement and data collection should be made on the basis of a more restricted population subset with an appropriate operating profile and physical environment. The important thing is to identify a **representative** population.

The standard makes the point that for field devices, experience may be based on both safety and non-safety applications (e.g. control and monitoring) where the duty is essentially comparable.

Clearly equipment that is brand new to market cannot be considered eligible on the basis of prior-use.

The standard requires that dangerous failure rates should be identified with at least 70% confidence. An estimate of dangerous failure rate may be made from the number of dangerous failures that have occurred in an identified population of devices by using the Poisson distribution.

As an algebraic approximation, the following equation may be used:

$$\lambda_d \approx \left[ 0.16 + \left( (1.065 + F_d) \times \frac{128.6}{E} \right) \right] / 114.2$$

Where:

$\lambda_d$ is the dangerous failure rate (Fails/Year) with 70% confidence

$F_d$ is the number of dangerous failures experienced

$E$ is the operational experience (Unit-Years)

An indication of a range of values is given in the table overleaf.

Note: Failure rates of less than $0.16/10^6$ hours (MTBF>713 yrs) are not supported by this approximation.

## CERTIFICATION & PRIOR-USE

| Number of dangerous failures | Unit-Years of operation | MTBF$_d$ (Years) |
|---|---|---|
| 0 | 10 | 8 |
| 0 | 30 | 24 |
| 0 | 100 | 75 |
| 0 | 300 | 185 |
| 0 | 1000 | 384 |
| 1 | 10 | 4 |
| 1 | 30 | 13 |
| 1 | 100 | 41 |
| 1 | 300 | 109 |
| 1 | 1000 | 268 |
| 2 | 10 | 3 |
| 2 | 30 | 9 |
| 2 | 100 | 28 |
| 2 | 300 | 77 |
| 2 | 1000 | 206 |

**Table 13.1** Poisson Approximation values

[A case might be made for simply dividing the number of failures by the total period (operational experience) to identify the failure rate, but this is mathematically optimistic and we could hardly claim with zero failures experienced that the failure rate is zero!]

Safe Failure Fraction (SFF) may be based on an evaluation of the nature of the equipment and its deployment; if it is good quality, industrial grade equipment, configured to be failsafe, it would not seem unreasonable to assume that the dominant failure mode is safe and to nominate a safe failure fraction in the 60-90% range. On this basis a working assumption of a safe failure fraction of 75% would seem reasonable. If there is some particular design concern that militates against safe failure you may judge safe failure fraction to be NOT dominant and adopt the corresponding hardware fault tolerance.

Note that contrary to the assumptions in some certificates, partial stroke testing is **not** a diagnostic; it is a partial proof test and therefore does not improve the SFF. (To qualify as a diagnostic the stipulation is that *'the diagnostic test interval of any subsystem...shall be such that the sum of* *the diagnostic test interval and the time to perform the repair of a detected failure is less than the Mean Time To Restore (MTTR) used in the calculation to determine the achieved safety integrity for that safety function'.* Since partial stroke testing is typically used at a relatively low frequency (with a period of days to months) the implication is that if PST were to be claimed as a diagnostic provision, the MTTR for detected failures would be correspondingly large. This makes for some difficulty in comparing certified offerings from vendors; the rigorously correct claim from one vendor may be seen as disadvantaged in comparison with the incorrect claim for improved SFF from a second vendor for items with identical reliability performance. It should be remembered that functional safety is not simply a numbers game (although many try to play it that way) and there is more to equipment selection than the raw, unqualified comparison of the numbers on certificates.

If you have explicit records of safe and dangerous failures within a representative population, the more conservative value of the apparent SFF or the SFF estimated from the above approach might be used.

In terms of the rigour with which failure rate data is identified, the following may be typical possibilities:

**Low Rigour**
Estimate of population size and historical number of failures (or rather frequency of failure) within an identified process operation, using judgement from experienced personnel associated with that operation. Assume at least half the failures are dangerous, and calculate $\lambda_d$ as above. Determine whether or not safe failures dominate on the basis of consideration of the nature of the equipment as explained above. (If you have reason to believe the SFF is significantly less than 50% then assume the corresponding proportion of failures is dangerous when calculating $\lambda_d$.)

E.g. If a population of 30 units used for 10 years, with a failure judged to occur once every other year, total failures ~ 5, so assume $F_d$ = 2.5, E = 30 x 10 = 300, giving $\lambda_d$ = 0.015 fails/year.

**Intermediate Rigour**

Records of failures (but not whether safe/dangerous) within a representative population. Identify the number of dangerous failures as the total number of failures multiplied by (1-SFF)/100 [using SFF estimated as above] and calculate the corresponding $\lambda_d$.

E.g. If a population of 10 units used for 15 years, with 4 failures recorded; SFF assessed as 75%, so $F_d$ = 1, E = 10 x 15 = 150, giving $\lambda_d$ = 0.017 fails/year.

**High Rigour**

Explicit records of dangerous and safe failures within a representative population. Use the identified number of dangerous failures in the population to calculate $\lambda_d$. Identify whether or not safe failures are dominant on the basis of the more conservative value; apparent SFF or the estimated SFF from consideration of the nature of the equipment.

E.g. If a population of 50 units used for 12 years, with 14 dangerous failures recorded; so $F_d$ = 14, E = 100 x 12 = 1200, giving $\lambda_d$ = 0.008 fails/year.

However you estimate and nominate the failure rate and safe failure fraction of the equipment, you should establish a means of tracking failures through your maintenance records so that you can validate and revise if appropriate the figures adopted. (See Chapter 21.)

I can imagine howls of protest from some about the lack of rigour here, but the requirement is not to establish the equipment performance with absolute rigour or assurance; the requirement is to identify appropriate evidence. The important thing is to be able to demonstrate that you have thought about these issues sensibly and have not simply selected equipment on the basis of lowest price, or familiarity with the vendor, or because you liked the colour.

For field devices, the standard also recognises that users' approved equipment lists, which carry implicit evaluation of equipment suitability, may be used to support claims of operational experience, provided that:

- the list is updated and monitored regularly;

- field devices are only added when sufficient operating experience has been obtained;

- field devices are removed when they show a history of not performing in a satisfactory manner;

- the process application (duty) is included in the list where relevant.

IEC 61511 calls for additional assessment for some equipment at certain SILs as summarised in the table below. The scope of the assessment required is detailed in the standard and I will not duplicate that guidance here; the assessment need not be that arduous but an offhand approach would not be acceptable; a responsible, considered, recorded evaluation is required, with due consideration to the guidance given in the standard. Understandably requirements for SIL 3 are more onerous and are likely to represent a particularly significant challenge.

The standards recognise different categories of software:

- Fixed Programmable Language (FPL) e.g. as used in configurable instruments.

- Limited Variability Language (LVL) e.g. ladder logic and sequential function charts as deployed in PLCs.

- Full Variability Language (FVL) e.g. Assembler, C, Java etc. as deployed on general purpose computers.

The IEC 61511 requirements in respect of PU may be summarised as follows:

## CERTIFICATION & PRIOR-USE

| Component Type | SIL1 | SIL2 | SIL3 |
|---|---|---|---|
| **Non Programmable** | Evidence of suitability | Evidence of suitability | Evidence of suitability |
| **FPL** | Evidence of suitability + consideration of: unused features, I/O characteristics, modes of use, functions and configurations used, previous use | | As SIL2 + assessment in accordance with 11.5.4.4 |
| **LVL** | As FPL + operational experience as appropriate for SIL and component complexity and functionality | | Refer IEC 61508 |
| **LVL safety configured PE logic solver (General purpose PLC configured for safety application)** | As LVL + understanding of unsafe failure modes, configuration techniques that address failure modes, protection against unauthorised or unintended modification | As LVL + formal assessment in accordance with 11.5.5.6 | N/A |
| **FVL** | Refer IEC 61508 | Refer IEC 61508 | Refer IEC 61508 |

**Table 13.2** IEC 61511 2Ed. Prior-Use Requirements

Note that the standard distinguishes between general purpose PLCs configured for deployment on safety applications and 'Safety PLCs' which are specifically built for safety applications and certified in terms of their design and configuration and operating system/software development.

It would be nonsense to deploy certified equipment 'just to be safe' if existing equipment is well proven with an established spares holding and with craftsmen trained and experienced in their installation and maintenance. The introduction of certified but novel equipment might look good 'on paper' but could well degrade the integrity of the installation. Take heart; prior-use is a practical and often preferable option.

Certification has its place also and can be a useful way of substantiating capability. Reputable companies will go to some pains to do the decent thing here, but all sorts of games do get played. It is well to understand on what basis certification is offered and not simply accept certificates at face value.

## 14. SYSTEMATIC FAILURES

In IEC 61508:2010 Part 4, Systematic failure is defined as 'failure, related in a deterministic way to a certain cause, which can only be eliminated by a modification of the design or of the manufacturing process, operational procedures, documentation or other relevant factors'. The standard distinguishes these from random hardware failures which it identifies as more susceptible to statistical analysis. Probability of Failure on Demand (PFD) calculations are routinely performed using estimates of undetected dangerous random hardware failure rates, but this assumes the systematic failure possibilities have been addressed by other means. It might be argued that some systematic failures are susceptible to assessment using statistical techniques in the same way as random hardware failures and in Chapter 20 the case is made for assuming an additional failure rate associated with the specifics of the equipment deployment (rather than inherent to the equipment itself).

Although much is made of PFD calculations and supporting equipment certification, the rigour with which these concerns is often pursued may be of questionable value when considered in the wider context including the systematic failure possibilities. The SINTEF reliability data handbook [1] makes the point that 'vendor estimates of dangerous undetected failure rate are often 'an order of magnitude (or even lower) than that reported in generic data handbooks'. In their estimates of failure rate they identify a parameter r as the fraction of dangerous undetected failure rate arising from random hardware failures. Values of r for field equipment (sensors and final elements) typically vary between 30-50%. It can be seen that systematic failures may be very significant and may well dominate.  Even equipment that has been certified as Safety Integrity Level (SIL) capable and installed, operated and maintained in accordance with safety manual stipulations may be found to remain susceptible to systematic failures.  This chapter considers how these concerns may be addressed.

**Types of Systematic Failures**
We may identify different types of systematic failures, for example:

a) Hardware issues that are 'engineered in' as part of the design or installation

b) Software issues that are 'engineered in' as part of the design

c) Those that develop progressively e.g. impulse line blockage

d) Those that arise under certain operating or environmental conditions

e) Those that arise from human error e.g. erroneous modification/adjustment,

Of these, types a) and b) should be mostly captured during function verification & validation.  Types c) through e) are potentially more difficult to address but may be captured by appropriate testing techniques.

We may further categorise systematic failures as 'persistent' or 'transient':

Persistent failures will remain until rectified and will therefore be discovered by effective testing techniques. Transient failures would only be present whilst certain conditions exist; when the conditions change the failure will 'repair' itself. These are much the most difficult to capture.

Although the relationship between testing and random hardware failure rate is well understood, the usefulness of testing in capturing systematic failures is often overlooked – reliance is placed instead on employing equipment which has an appropriate 'systematic capability' i.e. possessing appropriate defences against systematic failures.

**Probability of Coincidence**

Although we typically cannot assign a failure rate to systematic failures and cannot therefore calculate a corresponding PFD in the conventional manner, we can identify an upper bound to the probability of the coincidence of a trip demand with a persisting systematic failure, as a function of the demand rate (D) and a systematic failure test interval (TS): Consider a demand arising once per year; if we test 10 times a year then we would expect to discover (and eliminate) any persisting systematic failure that arises within 0.1 of a year. The probability of a demand arising within a given test interval is 0.1 and on average would be half way through an interval and so we may assess that the probability of coincidence (Pc) of a demand with a persisting systematic failure must be less than 0.1/2 = 0.05. Generally:

$$P_C \leq \frac{1}{2} D . T_s$$

Demand rate here corresponds with the frequency with which the hazard would be realised without the safety instrumented function. That is, with other provisions such as conditional modifiers or mechanical relief accounted for (the 'intermediate likelihood frequency' that is identified in Layer of Protection Analysis) and in the process sector will typically be very much less than 1/year. (See Chapter 6.)

**Systematic Failure Testing**

The systematic failure test interval identified above is that associated with a completely comprehensive test i.e. one that would identify any systematic failures.  Note that normal proof testing, which is predicated on random hardware failure rate, and as typically undertaken during process shutdowns, will not usually provide complete coverage of the systematic failure possibilities.  Note in particular that some obscure failure mechanisms may present under normal operating conditions that may not be apparent with proof test performed during plant shutdown. Examples:

- A valve that strokes satisfactorily during plant shutdown may not be effective in stopping full flow when pipework is stressed under operating pressure and temperature.

- A test button on a tuning fork would not identify if the wrong length probe had been installed in the top of a vessel.

- A workshop calibration check on a pressure transmitter would not reveal problems with the field power supply or impulse line.

- Signal injection would not reveal a ranging error on a transmitter.

- Line vibration may interfere with no/low flow detection by a vortex meter.

For systematic failure testing, the best testing approach is to drive the process to the trip point, but this may be impracticable or potentially hazardous.  If it is a practicable option, critical consideration must be given to the possibility that the function fails and the process may be driven to a potentially unsafe state; appropriate procedures and back stop provisions must be in place.

Other than driving the process to the trip point, coverage can typically be enhanced by in-situ testing with minimal disturbance to the installation and with elements exercised during plant operation (rather than at shutdown).  It may be possible to corroborate measurements through cross checks whilst the plant is operating.  The better practice is to avoid interrupting power supplies since this would cause equipment to reinitialise and possibly 'repair' systematic software failures that would otherwise be revealed.

Note that real trip events constitute the best possible test; if the necessary data can be captured to demonstrate that a trip performed satisfactorily then this would allow the next systematic failure test to be deferred. Note that this is not simply a matter of observing that the plant did shutdown safely; you would need to demonstrate that all the function elements performed as per design. For example; did the flow stop because the emergency shutdown valve closed on command from the safety instrumented system or was it because the control valve closed on command from the basic process control system? Note also that elements that are employed by multiple functions do not need to be tested for each function; recognition of this can massively reduce the testing burden.

One possible response to these considerations would be to specifically test for systematic failures frequently enough so that the $P_C$ contribution is suitably constrained. It may be that this is not a practicable option for many functions, but the approach may be worthy of consideration.

**Assessing Systematic Integrity**

The approach prescribed in IEC 61508 for establishing defences against systematic failures is qualitative rather than quantitative and there are numerous tables listing techniques and measures that may be employed in supporting claims of systematic capability (Part 2 Annexes A and B). These techniques and measures are called upon to be employed in a variety of combinations (some measures are held to be replaceable by alternatives), with differing degrees of effectiveness (High/Medium/Low) and differing degrees of recommendation (Mandatory/Highly Recommended/Recommended) depending on the SIL to be supported. Although not explicitly declared, the notion is that the measures and their effectiveness can be suitably weighted to allow an evaluation of their aggregate contribution. This is a somewhat convoluted approach in which the apparent 'rigour' is largely illusory; for example; 'The effort required for medium effectiveness lies somewhere between that specified for low and high effectiveness'. In practice it would be difficult for an end user to assess an element against the requirements for techniques and measures; more typically this would be for the manufacturer.

**Prior-use Assessment**
Although not explicitly stated, the requirement for systematic capability is implicit in the requirements for 'prior-use' evaluations. These include 'evidence of suitability', which includes:

- consideration of the manufacturer's quality, management and configuration management systems;

- adequate identification and specification of the components and subsystems;

- demonstration of the performance of the components and subsystems in similar operating profiles and physical environments;

- the volume of the operating experience.

The maturity of products is critical to the assessment, it being through feedback from experience that confidence in deployment can be gained. It may well be that manufacturer's experience leads to modifications, particularly to firmware, and it is critical that these modifications are managed with appropriate quality assurance measures. The following questions are suggested as an appropriate outline for an assessment of systematic capability of an identified element type and build; appropriate thresholds could be identified against which the level of systematic capability might be nominated or disallowed. In assessing a product it then becomes a question of the duty holder satisfying himself in regard to these characteristics and placing the assessment on record.

1. For how long has the element been manufactured? (Relates to manufacturer's experience.)

2. For how long has the manufacturer been established in the process sector? (Does the manufacturer have an established reputation to protect?)

3. Is the element series type manufactured in volume? (If not, manufacturer's confidence from customer feedback will be limited.)

4. Has the manufacturer operated an accredited QA scheme covering manufacture and development of the element, including any software?

5. Are there known issues with reliability? (Would immediately disqualify a product unless the issues can be eliminated by specific deployment measures.)

6. What type of software language is employed: Fixed Program, Limited Variability, Full Variability? (Refer IEC 61511 Section 11.5)

7. Can any programmable capability be secured?

8. Are comprehensive installation and operating instructions available?

9. Are comprehensive specifications available? (Functional, Operating & Storage Conditions, Performance, Physical, Electrical etc.)

This assessment to be performed by a responsible and competent engineer and held as an archive record in support of the judgement of the systematic capability of the element.

### Legacy Installations

This is all well and good when procuring a **new** element on the basis of an existing element, (perhaps to add a similar function or replace an existing element with the current version of the same model), but would be problematical if looking to make an assessment of an **existing** (legacy) in-service element to substantiate its continuing deployment in fulfilment of a retrospectively nominated SIL. Difficulties would likely arise in acquiring supporting quality assurance information, (particularly if the in-service model has been superseded) since we would need to consider the history of the manufacturing systems **before** original procurement, rather than their current status. It might also be the case that the manufacturing history prior to original procurement is not extensive; an element may have been in service for 10 years and manufactured for little more than this period. However, if the user experience gives no reason to doubt the suitability of the legacy equipment, it would be perverse to insist on its replacement with something that is not proven on that particular duty. We might, then, accept extensive and satisfactory experience (with a record of time in service and details of the specific duty, environment and service) in lieu of 'consideration of the manufacturer's quality, management and configuration management systems.'

Since our concern here is with defences against systematic failure, we do not need to identify the performance of a large population for the purposes of our legacy assessment; an assessment might be made against a single instance. (The performance of an identified population would pertain rather to the random hardware failure rate assumed.) We would, however, need to consider the in-service performance of the element in question and consider whether there have there been any unresolved issues with reliability due to systematic concerns. The longer an element has been in service the more opportunity there will have been for any systematic weakness to manifest itself and the greater our confidence will be that an element has adequate defences against systematic failures. It would be necessary to be satisfied that the element has responded to a real demand or has in some other manner been comprehensively tested for systematic failures. It would otherwise be conceivable that the element may have suffered an unrevealed systematic failure. It should be understood that this assessment would only be valid for that specific legacy duty.

This assessment process would establish the Systematic Capability of an element for the identified duties; it would remain to compile a full safety specification for the element to allow evaluation of requirements in respect of hardware fault tolerance, failure rate and stipulations for installation, configuration, maintenance and operation. This may be prepared by consideration of the element nature and design and references to the manufacturer's product specifications and manuals, together with estimates of failure rates from manufacturer's data, generic databases or experience with identified populations with circumstances of deployment suitably close to that proposed.

## Conclusion

It should be recognised that absolute guarantees of fitness-for-purpose are simply not available; we seek rather a degree of confidence. A responsible and critical (but relatively straightforward) examination of the provenance and in-service history of elements may provide this confidence. We should recognise also that the ultimate safety performance of elements is often governed by the specifics of the equipment deployment and the maintenance management provisions rather than the intrinsic characteristics of the equipment itself.

## References

1. Reliability Data for Safety Instrumented Systems, PDS Data Handbook 2010 Edition, SINTEF

## 15. LEGACY PLANT

If you have a legacy process plant, i.e. one designed without consideration of the requirements of IEC 61508/61511, the question arises; how do you now address this? This chapter outlines a practicable approach which will help identify and address any critical shortfalls in your protection provisions when measured against the benchmark of IEC 61511.

Understand from the outset that absolute compliance is a fantasy promoted by the misguided or the unscrupulous; they would have you 'rip-it-all-out-and-start-again', but that is not typically a practicable approach. The requirement is rather to pursue compliance 'so far as is reasonably practicable'. Recognise that this will require the exercise of professional engineering judgement about what is appropriate in the context of your operation; there is unlikely to be just one definitive solution.

**SIL Determination**
The first requirement is to identify just what the safety instrumented functions are, not necessarily as straightforward an exercise as might first be imagined (see Chapter 5).

Once you know what the safety instrumented functions are, your next mission is to perform a Safety Integrity Level (SIL) determination exercise for each of these functions. This may be through a variety of techniques e.g. risk matrix/risk graph/layer of protection analysis/fault tree studies. (See Chapter 7 for further details on these approaches)

This does need some care however; over ambitious targets for tolerable risk, successive compounding of undue conservatism in the evaluation, or overlooking of significant conditional modifiers can quickly lead to inflated SIL targets. This can lead to unwarranted difficulty with the engineering and the spurious diversion of resources from where they may be more usefully deployed. Conversely, improper application can give misleading under-estimates of the required risk reduction and the corresponding SIL targets.

If your installation was originally designed and built in accordance with established industrial good practice, it is likely to be largely compliant. There may be some minor deficiencies that can be reasonably straightforwardly addressed, but a wholesale re-engineering exercise is unlikely to be warranted. Typically most of the protection functions will not be SIL rated, several may be SIL1; you might occasionally meet SIL2, exceptionally SIL3. If you meet SIL4, the likelihood is that you are doing it wrong or that your process design is seriously flawed. Historical good practice design provisions will typically be good for up to SIL2. SIL3 is a game changer, and is likely to require a degree of redundancy in the protection function. If your SIL determination process identifies a large number of SIL rated functions or a surprising number of higher SILs, it may well be that the determination approach is flawed or inappropriately calibrated.

Conversely, it may be that the SIL determination process confirms a higher SIL than was perhaps anticipated; this is typically where the process offers most value - when it reveals a hitherto unrecognised significance to a protection function. If you cannot square the results of your SIL determination exercise with your judgement, you need to look harder to discover from where the apparent anomaly arises, and refine either your analysis or your judgement (or both).

Those safety functions that are not SIL rated fall outside the remit of IEC 61511 and the requirement is simply that they be maintained, tested and inspected in accordance with good practice, albeit without any specific performance target and without the same rigour in terms of supporting documentation and design verification. What you cannot legitimately do is regard these functions as 'fit-and-forget'. (See Chapter 16)

So now you know what protection functions you have and what their respective SIL targets are.

**Safety Requirements Specification**

You are now in a position to write a Safety Requirements Specification (SRS) for the SIL rated functions. If you turn to IEC 61511/61508 you will find a daunting list of things that should be addressed by an SRS, but this is typically to inform an appropriate design and procurement exercise when starting with the proverbial 'clean piece of paper'; you already have your design, procurement is a distant memory. A 'retrospective' SRS for a legacy installation is essentially about defining just what the safety instrumented function is, and its performance requirements; i.e. what SIL target is identified, what is the process safety time, is tight shut-off required, what is the trip setting, and what operational features are required i.e. resets and overrides. (Note that tight shut-off is typically not needed for safety functions; effective throttling of flow is usually sufficient to supress the hazard.) This is a much reduced subset and should not be difficult to capture as a formal record. The key thing is that the SRS should support your management of change process, so that any modification proposal can be assessed against the identified functional requirements. If your review identifies a shortfall in your SIF provision, then the SRS you generate should reflect the target requirements, not just the 'as-found' circumstances.

**Verification**

The next step is to perform a verification of the design for the SIL rated functions; this essentially means;

    a) Undertaking a calculation of the Probability of Failure on Demand

(PFD) of the individual functions to confirm compliance with the SIL target, and

b) Confirming that the equipment deployed on the safety function is fit-for-purpose and supports the function architecture (see Chapter 14).

If the PFD is too large for the target SIL, your options are;

- Revisit the calculation (is the failure rate data too conservative?)
- Use more reliable equipment (with improved failure rates)
- Enhance the function design (change the architecture)
- Test more often

If the equipment is found not to be 'fit-for-purpose' your options are:

- Replace the equipment (use better equipment)
- Increase the diversity in the architecture (use different equipment in redundant branches)

In order to support the PFD calculation you need to identify just what the architecture of each function is and what equipment types are deployed. A simple block diagram is usually helpful. The PFD calculations will confirm how often you will need to proof test the SIL rated functions in order to meet the SIL target for PFD.

## ALARP Provisions

If, in order to meet the requirements of the standard, the revisions to the design provisions would require what are considered to be unwarranted difficulty and expense, it may be possible to demonstrate that the existing provisions already reduce the risk to be 'As Low as Reasonably Practicable'. This would require a demonstration that the cost of the improvement would incur expense that is 'grossly disproportionate' to the value of the postulated decrement in risk. This is an entirely legitimate approach that may avoid the unwarranted diversion of resources; it is not to be invoked on a whim however, merely in an attempt to avoid inconvenience.

**Proof Testing**
Having identified how often functions should be tested, you need then to establish an appropriate proof test and inspection regime to maintain the required PFD of the functions and identify and address any significant deterioration that may occur (a function may still perform even though clearly suffering distress). Appropriate proof test procedures need to be written. These should be perfectly explicit and function specific, detailing every step and appropriate pass/fail criteria. Highly generic, high level procedures e.g. 'calibrate and test pressure trip' are not acceptable. (See Chapter 25)

**Competence**
Technicians engaged in the maintenance of SIL rated functions need to be suitably competent for that purpose. That is not to say that they all need to be qualified functional safety practitioners, but they need an awareness of SIL/61511 and an understanding of the implications for repair/maintenance and change control. Given that they are already generally competent in maintenance of process instrumentation, the training requirements are relatively modest; typically a half-day awareness training course would be sufficient. (See Chapter 12)

**Monitoring**
Finally (!) you need to establish a mechanism for monitoring the ongoing performance of your SIL rated functions (See Chapter 29) and managing equipment obsolescence and ageing beyond useful life. A simple database identifying what equipment is deployed and where, and what its age is, will be helpful. (See Chapter 21)

**Conclusion**
For the most part, installations that have been designed in accordance with historical good practice will be largely compliant with the provisions of IEC 61511. Good practice provisions will typically be good for up to

SIL2. Likely deficiencies will be; identification of specifically what constitutes the SIF, formal verification evaluation of the design, and explicit proof testing procedures. Typically the requirement against legacy installations would not be one of re-engineering, but rather of establishing appropriate documentation and management provisions to ensure ongoing fitness-for-purpose.

## 16. LOW INTEGRITY FUNCTIONS

The Health and Safety Executive (HSE) have published guidance for their inspectors on 'Management of instrumented systems providing safety functions of low / undefined safety integrity' (Operational Guide OG46). The target audience for the guidance is the HSE's specialist inspectors, but the guidance is openly published to help inform a wider audience. The end user looking for specific guidance on fulfilment of his duties might well be disappointed however; the guidance is relatively high level and generic. The provisions it identifies are clearly related to those promulgated by the functional safety standards IEC 61508/61511 for Safety Integrity Level (SIL) rated Safety Instrumented Functions (SIF), but there can be no expectation of full design verification or indeed of implementation of the wider safety lifecycle, otherwise the guidance would simply have invented SILO! The document identifies eight provisions that low integrity SIF should be subject to; let us consider how the duty holder might respond to each of the required provisions:

> *"the persons who have responsibilities for the instrumented system shall be suitably competent"*

I suggest no extraordinary provisions are called for here; the basic criteria and processes for recruitment and training of instrument personnel to allow their effective deployment on any instrumented facility should be adequate, there being no exceptional requirement for low integrity SIF.

> *"clear, precise and unambiguous specification of the safety function"*

This might be as simple as an indication on a Cause & Effect matrix; as long as the key functional requirement is identified. The specifics of the implementation might otherwise be handled as a matter of recognised good practice and established site standards and not necessarily tied to specific safety system provisions that might well be stipulated for SIL rated SIF.

> *"independence between control and safety functions wherever reasonably practicable"*

This is potentially confusing since it might well be the lack of independence that is the reason for the low integrity. If the overall risk is already ALARP then there would be no case for engineering enhanced independence and the requirement on other defences may well have already been identified on the basis of the limited risk reduction from the low integrity SIF. Note that an additional trip initiation sourced from a Basic Process Control System (BPCS) sensor might be a useful enhancement of overall integrity – despite the lack of independence.

The critical consideration is that we should not be relying solely on a non-independent SIF to protect against failure of the control function; a failure might well disable the SIF at the very time it is needed to protect against the control failure!

> *"accurate, accessible, controlled and easily understood engineering documentation showing the component parts of the instrumented system and how they are configured. Examples of engineering documentation include loop or circuit diagrams, equipment data sheets and records of parameter settings"*

A matter of good practice for any installation, not just the SIF, whatever their integrity. No exceptional requirement here.

> "periodic inspection of the instrumented system, for example visual or more detailed inspection to reveal evidence of deterioration or unexpected modifications"

This is entirely appropriate and would be in the interests of plant availability as well as safety – the task should not be unduly burdensome; a simple inspection tour to confirm there is no obvious degradation might be simply included as part of the proof test provision.

> "periodic maintenance of the instrumented system in line with manufacturers' recommendations and general good practice periodic maintenance of the instrumented system"

Examples include calibration, cleaning or flushing. Again this may well be prudent for reasons of availability (avoidance of spurious operation) and nothing beyond the normal good practice provisions that are adopted for the rest of the plant would be needed. Rather than actual calibration via a reference, where practicable, simple cross checks between readings from different indications might be considered sufficient to confirm the satisfactory condition of a low integrity measurement.

> "periodic testing of the instrumented system, at intervals defined by suitably competent persons, for the purpose of revealing dangerous undetected faults"

This is a potentially difficult area, but if the SIF, albeit low integrity, is claimed as part of the management of risk provisions it cannot be regarded as 'fit-and-forget'. That said, the testing should not be unduly burdensome; it should, of course, be commensurate with the integrity claim. For a SIL rated SIF this requirement would be established through a Probability of Failure on Demand (PFD) calculation, but, as mentioned in the introduction, there can be no expectation of the full design verification for a low integrity SIF. We can however establish some ranging shots. If a low integrity SIF was implemented in the BPCS the highest Risk Reduction Factor (RRF) we could claim is 10 (PFD of 0.1). The lowest failure rate we could claim is 1 in 100,000 hrs (nominally 1 in 10 years). Since for a 1oo1 function PFD = ½ x failure rate x test interval, the implication is that the test interval should be 2 years. This then is our starting point. A failure rate of 0.1/year is quite modest and for a low integrity SIF where there is degree of confidence that a failure rate of better than this is to be expected, (on a benign duty, independent of the BPCS say), I suggest it would not be unreasonable to increase the test interval to 4 years, say. These then are possible default values for test interval. (If there were concerns that failure rate might be greater than the nominal 0.1/year there would of course be a corresponding requirement to reduce the test interval.) If a full coverage test would be problematical, then by performing a partial coverage test at increased frequency, the requirement for a full test may be legitimately extended and the overall PFD maintained. Consider a nominal 50% coverage test (sensor + logic or logic + final element say, assuming sensor and final element have dangerous undetected failure rates of the same order). If we double the test frequency, so once a year or once every 2 years for the examples above, then the full 100% test or overhaul could be performed at a corresponding 3 or 6 years. If a risk reduction of less than 10 is claimed there would be a corresponding increase in these test intervals. (See Chapter 25)

> *"management of change, including control of access to software functions and backing up of software-based systems"*

Again this may well be prudent for reasons of availability (avoidance of spurious shutdown) and nothing beyond the normal good practice provisions that are adopted for the rest of the plant would be needed.

From these considerations, it can be seen that nothing beyond the normal good practice provisions for NON-safety functions is typically required; OG46 can be seen as simply a requirement from the regulator to extend these practices to low integrity SIF. If you choose not to implement these practices on your non-safety functions that is your affair, but the inspector might well be prompted to wonder whether the overall management culture was appropriate to the maintenance of a potentially hazardous facility.

Many low integrity SIF are likely to be alarm or control functions. If a specific risk reduction claim is made for them (typically a factor 10) it is suggested they should be proof tested in accordance with the guidance offered above. If no explicit risk reduction is claimed, then as a proportionate provision a simple health check (rather than a full proof test) may be considered sufficient; no obvious sign of degradation, equipment powered, no fault flags, signal level and variation consistent with operating conditions, lamp test etc.

One other consideration not covered in OG46; if the equipment deployed on low integrity SIF (or indeed other control or monitoring functions) is also deployed on high integrity SIL rated SIF, the broader equipment population (including the low integrity and control functions) should be included in the ongoing monitoring of equipment reliability in order to validate the assumptions concerning failure rate for the SIL SIF equipment; the broader total population will provide greater confidence in the failure data. (See Chapter 21.)

## 17. VERIFICATION & VALIDATION

Easily confused! And the terms are often used interchangeably, but to be nice (and correct) about it in terms of the standards:

**Verification**: confirmation that the outputs of each individual safety lifecycle phase are consistent with the inputs, objectives and requirements of that phase.

**Validation**: confirmation that that the SIF(s) and SIS after installation meet the SRS in all respects.

The standard requires that both are performed by 'examination and provision of objective evidence'.

What does this mean for real world projects? Essentially verification would be applied to each individual project deliverable; each specification, each drawing, each schedule. These all constitute 'output' deliverables that become the starting point or data input for downstream phases. The requirement is to have all such deliverables critically reviewed before formal, final issue. The simple means of implementing this is to employ the familiar check and approval protocols typically used in project engineering. That said, it must be acknowledged that the check and approval protocols often employed in projects don't always amount to much; they might well be the most cursory of reviews and they may add little value. Approvals in particular often achieve little beyond slowing the project with a management bottleneck, and enforcing acknowledgement of the management hierarchy. (Approval is often, by default, limited to departmental heads, regardless of whether or not this is sensible; this is perhaps more of a political statement than a technical one.) In functional safety (our particular concern, but also in many other areas), this will not do. If you are to use check and approval as your verification mechanism then the obligations on the checkers and approvers need to be explicitly declared and understood. This needs real care and it would typically be useful to have different grades of checking and approval that could be nominated depending on the nature of the deliverable and the degree of rigour required.

We can anticipate a spectrum from complete independent production with discrepancy checking, through comprehensive checking on a sample basis, to simple overview of format and content.

Whereas verification is of each individual phase, validation is typically not implemented as a discrete standalone activity, rather it is achieved through a suite of installation and commissioning checks e.g. point-to-point wiring checks, as-building, insulation tests, valve stroke tests, system site acceptance testing, function tests etc. and there is nothing particularly new in any of this for a well-executed project. The key point for SIL rated functions is that all of these must be done with due rigour and traceability by competent people, with acceptance criteria that can be traced backed to the safety requirements specification.

# 18. FUNCTIONAL SAFETY ASSESSMENTS (& AUDITS)

In respect of Functional Safety Assessments (FSA), the IEC 61511 standard calls for 'A procedure ...defined and executed...in such a way that a judgement can be made as to the functional safety and safety integrity achieved by the safety instrumented system. (Part 1, Clause 5.2.6.1.1)

The standard calls for an FSA at up to 5 stages:

Stage 1: After SRS development

Stage 2: After design

Stage 3: After installation, commissioning & validation.

Stage 4: During operation & maintenance

Stage 5: After modification

The standard makes FSA Stage 3 mandatory before the introduction of hazards, but the earlier stages are optional. You might well wonder then, why anyone would bother with these earlier stages. The point is that fixing problems at stage 3 can be awkward and expensive; if you can catch problems early on, so much the better. If however, the territory is familiar, and no significant issues are anticipated, it may be that stages 1 and/or 2 can be omitted without undue exposure.

FSA is a distinct activity from verification or validation and is concerned rather with an overview of the safety lifecycle processes and confirmation that they have been faithfully executed. This is demonstrated in IEC 61511 which identifies FSA scope, prior to the introduction of hazards (up to and including FSA3), as confirmation that:

- The hazard and risk assessment **has been carried out;**

- The recommendations arising from the hazard and risk assessment that apply to the safety instrumented system **have been implemented and resolved;**

- Project design procedures **are in place and have been properly implemented;**

- The recommendations arising from the previous functional safety assessment **have been resolved;**

- The safety instrumented system is designed, constructed and installed in accordance with the safety requirement specification, any differences **having been identified and resolved**;

- The safety, operating, maintenance and emergency procedures pertaining to the safety instrumented system **are in place;**

- The safety instrumented system validation planning is appropriate and the validation activities **have been completed;**

- The employee training **has been completed** and the appropriate information about the safety instrumented system **has been provided** to the maintenance and operating personnel;

- Plans or strategies for implementing further functional safety assessments **are in place.**

It can be seen that the emphasis throughout is on confirmation of implementation and completion.

## FUNCTIONAL SAFETY ASSESSMENTS (& AUDITS)

The extent and rigour of any assessment is a matter of judgement and is acknowledged by the standard to depend on the size and complexity of the project, the safety significance, previous experience of similar systems and standardisation of design features. For many projects an FSA need not be a particularly onerous exercise. The critical thing is that the person(s) performing the FSA should take a suitably detached view, for this reason the standard requires that the assessor(s) be independent from those actually executing the work. It is the detached view that is important, the independence is a means to that end. This is potentially fraught territory since FSAs can become politically charged; it should be understood that 'independent' is not synonymous with 'disinterested'. A party may well be independent but they may have an undeclared agenda of belittling the competition or of drumming up further business – of course the competition may deserve criticism if they are truly incompetent, and suitably discreet promotion of services might well be appropriate, but the spurious inflation of concerns (scaremongering) is not acceptable.

FSAs are represented as discrete milestone activities in the safety lifecycle, but that is an unrealistic ideal; in practice an FSA will be informed by on-going activity; project execution cannot be expected to stand still whilst an FSA is undertaken, and total completion of all appropriate phases ahead of an FSA would be an impracticable expectation - It would potentially delay pertinent findings that might have implications for following phases. (FSAs are rather like quantum mechanical 'particles' (wave-icles) the closer you look the more 'fuzzy' and 'spread out' they become!) Similarly, although the principle of '...before hazards are introduced' is sensible enough as an ideal, it may not be a practicable option. Of course, if there is any reason to doubt the proper functioning of a SIF, then it would be irresponsible to introduce the hazards, but if satisfied on this count, and the issue is rather one of degree, then it might be acceptable to perform an FSA stage 3 as soon as is practicable after start up. These are properly matters of professional engineering judgement from suitably competent practitioners.

Assessments should not simply identify shortfalls or failings, but should give an indication of their potential significance in the context of the project. Many minor deficiencies might be discovered, but that is not necessarily to say that the functional safety provisions will be significantly compromised. Let me be clear however; if the assessor has reason to believe the functional safety provisions will be significantly compromised he should declare this in unequivocal terms. Typically the deficiencies will be in traceability and quality assurance rather than the engineering *per se*; the design and implementation may be perfectly satisfactory, but the demonstration in terms of documentation and project protocols may not be evident. The assessment should be appropriately nuanced; we are looking for an informed appraisal, not a hatchet job!

The significance of FSA Stage 4 should not be overlooked; this is where you evaluate, in the light of experience and actual system performance, the validity of the assumptions made during the design phase (e.g. in respect of failure rates and demand rates). (See Chapter 21.)

Audits in respect of functional safety are no different in principle than any other quality assurance audit; it is simply a check on compliance with the procedures prescribed for the project.

# 19. ELEMENT METRICS: USEFUL LIFE, MISSION TIME, MTBF, MTTF, MRT, MTTR

**Useful Life**

Useful life is the period during which equipment is identified as having a constant failure rate, which is an assumption made in pfd calculations. It is classically the central period on the failure rate-time 'bathtub' curve that lies between the infantile failure period (where failure rate is relatively high and falling) and wear out (where failure rate is increasing). Some will tell you that after the useful life has expired you must replace the equipment; it isn't necessarily true. The reliability of equipment does not fall off a metaphorical cliff edge upon expiry of the useful life. The usual model of the wear out phase is of a progressive and accelerating climb in failure rate.

**Mission Time**

Mission Time is sometimes identified for equipment items and is the time after which equipment should be replaced (and is often used with this meaning in the machinery sector) or, if possible, otherwise inspected and overhauled to return the equipment to the 'as new' condition. It provides for a resetting to theoretical zero of the instantaneous pfd for the item, which would otherwise grow unacceptably because of possible undetected failures. This approach may be used where equipment is not periodically proof tested or where the testing provides only partial coverage, in which case there would be an accumulating residual probability of failure on demand. Note that this is subtly different from the useful life. In principle an item could have a useful life of 20 years say, (during which the failure rate is identified as constant), and a mission time of only 10 years after which its pfd contribution may have grown to the limit of acceptability.

### Mean Repair Time (MRT) & Mean Time To Restore (MTTR)

Easily confused and often used interchangeably, but not (quite) the same thing! MRT is the average time it takes to fix your SIF once your proof test has revealed that it is broken. MTTR is the average of the time taken for any diagnostic provision to discover the failure added to the time it takes to fix it. These are the meanings used in IEC 61508. You may elsewhere see MTTR used to abbreviate Mean Time To Repair (rather than Restore). The distinction is a subtle one; it is about calculating the total downtime associated with a failure. If the failure is found through a proof test, you will, on average, have to wait for half the proof test interval to find the fault; you then add the time required to fix the fault. If the fault is picked up through the diagnostics, it is the diagnostic test interval that counts. If there are no diagnostics, the MTTR is irrelevant. For many applications a default of MRT=MTTR is assumed, which is reasonable as long as the diagnostic interval is much smaller than the MRT. (If it will take days to fix, a diagnostic interval of minutes can be disregarded.)

Note also that if you shut down on detection of a fault (or don't allow start up following a proof test whilst shutdown until it is fixed), the effective MRT is zero. The calculation using MRT accounts for the possibility that the SIF (or parts of it) would not be available for the repair time, but if no demands can arise during this period the MRT becomes irrelevant.

### MTBF & MTTF

This is horribly confusing territory and the same terms are used to mean subtly different things in different contexts or industries.

Strictly, the Mean Time Between Failures, is the average time between failures, and is the reciprocal of failure rate where this is assumed to be constant (before wear out). This is the parameter that is used in the exponential distribution that is commonly used to model reliability of systems on the basis of a constant failure rate. Sometimes a distinction is made between repairable systems when the time it takes to fix it when it does fail (downtime) is included, and those that are non-repairable.

The MTTF (Mean Time To Failure) is the average time to failure including the wear out phase. You may see the term Mean Time BEFORE Failure used, which is really the same thing as MTTF, but the term is not used consistently.

The between/before, repairable/non repairable, distinctions are often not explicitly made and the terms are often used interchangeably. You may well see MTBF given as Mean Time BETWEEN Failures but actually meaning BEFORE (i.e. MTTF).

You may see MTTF/MTBeforeF rather than MTBetweenF declared for a non-repairable element or system, but it would be a mistake to use this in the PFD calculation for which we need the failure rate during the 'useful life' when a constant failure rate is assumed.

Since in our calculations we always need failure rate rather than MTBF, you might wonder why we should bother with MTBF at all. The nice thing about MTBF is that it provides a more intuitive metric; it is much easier to see the significance of an MTBF of 200 years rather than a failure rate of 0.005/yr. This does not mean that the component will last 200 years, rather it means that if you have 200 of them, you can expect, on average, one to fail every year. This allows an evaluation of experience and expectations.

It is entirely possible that a device will have an MTBetweenF of 200 years and an MTTF of 10 years.

# 20. ADJUSTING EQUIPMENT FAILURE RATES FOR DUTY

In order to calculate Probability of Failure on Demand (PFD) it is necessary to identify estimates of equipment failure rates. One approach is to rely on failure rate data declared in a Safety Integrity Level (SIL) certificate from a vendor or independent test house. These figures may however be optimistic in that they are typically for reference conditions which cannot necessarily be guaranteed to apply on an uninterrupted basis.

**Deployment effects**

In the 'real world', equipment may fail due to:

- Being stood upon
- Being dripped upon
- Scaffold strikes
- Lightning strikes/Voltage spikes
- Process excursions
- Process connection failures
- Compromised ingress protection
- Environmental extremes
- Inappropriate modification/adjustment

We might categorise these influences as being due to 'deployment'. They are likely to be difficult to quantify but may well dominate the actual failure rate of a function. It might be argued that these influences lead to systematic failures rather than random hardware failures, but although they may be systematic (in that they are '...related in a deterministic way to a certain cause' per IEC 61508), they may still arise on a random basis, (unlike other systematic causes such as errors in the safety requirements specification) and would be revealed by a proof test. Unless they are specifically 'designed out', any PFD calculation that does not make allowance for these concerns is likely to be unrealistic. You might well see equipment with certified Mean Time Between Failure (MTBF) figures of say 200 years with a safe failure fraction of 80%, implying that the MTBF for dangerous failures will be 800 years. This might well be an optimistic claim for 'real world' circumstances where deployment influences might undermine the certified failure rate.

The SINTEF reliability data handbook [1] makes the point that 'vendor estimates of dangerous undetected failure rate are often 'an order of magnitude (or even lower) than that reported in generic data handbooks'. In their estimates of failure rate they identify a parameter r as the fraction of dangerous undetected failure rate arising from random hardware failures. Values of r for field equipment (sensors and final elements) typically vary between 30-50%. It can be seen that systematic failures may be very significant and indeed will typically dominate. Although identified in the SINTEF handbook as a proportion of the overall dangerous undetected failure rate, which is not to say that the random hardware failures and systematic failures will be in a fixed ratio. I suggest that the contribution from systematic influences may be characterised as fixed for a given equipment type, installation and environment, and essentially independent of the inherent random hardware failure rate of the equipment.

Failure Mode Effects Analysis (FMEA) is often deployed as an evaluation tool as part of a certification process; it will examine the impact of component failures, but does **not** include any of the 'real world' influences listed above. The equipment is assumed to be operating at reference conditions. It is from these considerations that we might well come to have greater faith in reliability figures derived directly from plant history and associated judgements rather than declarations on certificates. What such judgements lack in rigour they make up for in being grounded in real experience. It could be argued that the real value of 'SIL' certificates lies in the assessment of systematic capability (SC) rather than of failure rate.

**Vulnerability categories**

An initial figure might be adopted where the manufacturer's certified figure for an equipment item is combined with default figures for deployment on the basis of 'vulnerability', i.e. susceptibility to additional factors. For these purposes the level of vulnerability may be assessed on the basis of three categories; environment, duty and exposure, as detailed in table 20.1.

## ADJUSTING EQUIPMENT FAILURE RATES FOR DUTY

| Vulnerability Category | Vulnerability Level |||
|---|---|---|---|
| | **Reduced** | **Standard** | **Increased** |
| Environment | Benign - well within capability, IP rating not critical to suitability, not subject to excursions beyond capability | Not having a 'Reduced' vulnerability in any ONE category; environment, duty or exposure. | Not having a 'Reduced' vulnerability in MORE than one category. |
| Duty | Benign - clean, non-aggressive (or not susceptible to fouling/attack) or not process wetted, little vibration, not subject to excursions beyond capability | | |
| Exposure | Limited - no exposed process connections/isolation points (e.g. impulse lines, press tx, valve manifold, capillary tubing) or protected from interference and inadvertent strikes, or remote from normal access. | | |

**Table 20.1** Vulnerability Level and Vulnerability Categories

Logic solvers and non-field equipment will typically be located in equipment/auxiliary rooms and not subject to the same range of influences as the plant sensors and final elements. They will however remain susceptible to unauthorised interference (probably well intentioned, but perhaps ill advised). Some equipment will be less susceptible to such interference and it is here suggested that vulnerability levels may be assigned on the basis of security together with an assessment of whether the assumption of benign duty and environment is valid.

|  | Vulnerability Level |  |  |
|---|---|---|---|
|  | **Reduced** | **Standard** | **Increased** |
| **Logic Solvers & Non-Field Equipment** | Inherently secure design (i.e. not dependent on added security control measures) e.g. Solid State Logic, Safety PLC<br><br>Benign environment - well within capability, IP rating not critical to suitability, not subject to environmental excursions<br><br>Benign duty - well within capability or derated | Secure Access<br><br>E.g. Relay System/Trip Amp in secure environment; locked cabinet/room authorised personnel access only.<br><br>Benign environment- well within capability, IP rating not critical to suitability, not subject to environmental excursions<br><br>Benign duty - well within capability or derated | Standard PLC OR Relay System/Trip Amp with unsecured access OR Non-benign environment OR Non-benign duty |

**Table 20.2** Vulnerability Level for Logic Solvers and Non-Field equipment

All sorts of qualifications might be introduced into the analysis of vulnerability, but given the levels of uncertainty in so many aspects of functional safety there is limited value in refining the analysis. The important thing here is to recognise that the 'raw' figure supplied by the vendor is not likely to be representative of the actual performance of the equipment in the field and it is prudent to make some allowance in recognition of the influence of deployment.

## Comparison of field and certificate data

As a means of establishing a suitable allowance we may compare nominal field values with vendor figures and identify the difference as being due to deployment. Table 20.3 compares some generic field values for field equipment, with nominal values representative of those typically found from vendor certification.

|  | Nominal MTBF$_d$ values (years) |  |  |
| --- | --- | --- | --- |
|  | Field Database | Vendor Certificate | Deployment |
| Press Tx | 150 | 1000 | 175 |
| Level Tx | 115 | 140 | 600 |
| Flow Tx | 225 | 900 | 300 |
| Temp Tx | 150 | 2900 | 160 |
| Remotely Operated Valve | 115 | 200 | 265 |
| Solenoid Valve | 125 | 200 | 340 |

**Table 20.3** Comparison of Field and Certified Values for Mean Time Between Dangerous Failures

From consideration of these 'ranging shots' and other database values, it is here suggested that a nominal figure of 300 years MTBF for dangerous failures be adopted in respect of 'standard' vulnerability deployment.

From inspection of a range of database values, and as a matter of judgement, a factor 3.0 shift in values for field equipment is here suggested in respect of different categories of vulnerability. It is acknowledged there is little enough science here, but having identified that equipment will be more or less vulnerable depending on the circumstances of the deployment, it is appropriate to make some corresponding allowance and the above is suggested as a starting point. Ultimately, the figures assumed should be validated by monitoring of the ongoing performance of the equipment. (See Chapter 21.)

Logic solvers and non-field equipment typically operate on benign duties in benign environments, but there will be some remaining vulnerability to the likes of voltage spikes, environmental control failures, electromagnetic and electrostatic effects, as well as unauthorised or misjudged interference. Given this residual vulnerability we may look for correspondence of the deployment element of standard vulnerability logic solvers and reduced vulnerability field equipment. In nominating the same value as for reduced vulnerability field equipment, the implication is that these residual influences represent 33% of the contribution to standard field equipment vulnerability, which does not appear unreasonable.

Suggested figures for dangerous failure rates due to deployment are given in the table below.

| Equipment Type | Vulnerability | Deployment MTBF$_d$ (years) | Failure Rate (years$^{-1}$) |
|---|---|---|---|
| Sensors | Reduced | 900 | 0.0011 |
| | Standard | 300 | 0.0033 |
| | Increased | 100 | 0.01 |
| Logic Solvers & Non Field Equipment | Reduced | 2700 | 0.00037 |
| | Standard | 900 | 0.0011 |
| | Increased | 300 | 0.0033 |
| Final Elements | Reduced | 900 | 0.0011 |
| | Standard | 300 | 0.0033 |
| | Increased | 100 | 0.01 |

**Table 20.4** Dangerous Failure Rates for Deployment.

The values in this table should be regarded as indicative rather than definitive. Users may wish to compile their own values, based on an evaluation of their specific applications' susceptibilities to 'real world' influences, particularly if site specific field data is available. A more comprehensive table might be compiled for individual equipment types, considering a range of generic database values and a range of vendor figures, but given the uncertainty in the data there may be limited value to be gained. (Functional safety is NOT an exact science; we seek to establish the right order of risk reduction commensurate with the circumstances.)

### Use of deployment figures

It is suggested that the 'raw' Safe Failure Fraction (as declared on the certificate) is assumed to apply to the deployed equipment unless there is specific evidence to the contrary.

So if we have a pressure transmitter with an MTBF (dangerous) of 750 years (failure rate 0.0013 yr$^{-1}$) on a 'standard' deployment, we would combine this figure with the default deployment failure rate of 0.0033 to give an overall figure of 0.0013 + 0.0033 = 0.0046 (217 years MTBF$_d$). With reduced vulnerability deployment the corresponding figure would be 0.0013 + 0.0011 = 0.0024 (417 years MTBF$_d$).

Note that a 'perfect' item of equipment would never have a 'real' (deployed) MTBF of better than the default deployment figure. A perfect pressure transmitter on standard deployment could not be claimed to offer an in-service figure better than 300 years $MTBF_d$. Where redundancy is claimed the usual approaches to common mode failures may be adopted. The intention here is to simply qualify the individual device failure rate figure.

This approach is proposed as a default; specific values might be identified for individual items on an exceptional basis. The implication being that the user should identify why any exception is considered appropriate.

As a 'sanity check' on the values and the approach, we may examine a notional final element sub-system consisting of a solenoid valve driver barrier, a solenoid valve, an actuator and remotely operated valve, with respective certified MTBF dangerous figures of 10,000, 200, 200, 200 years.

With standard deployment we have a sub-system $MTBF_d$ of 38 years:

$(0.0001 + 0.0011) + (0.005 + 0.0033) \times 3 = 0.0261/year.$

If we allocate 50% of the function PFD to this final element sub-system, we would need to test every 1.2 years to meet a mid SIL1 target for the function of 0.0316. If the vulnerability was 'reduced' the corresponding $MTBF_d$ would be 53 years:

$(0.0001 + 0.00037) + (0.005 + 0.0011) \times 3 = 0.0188/year$

with a corresponding test interval of 1.7 years.

For a pressure transmitter, repeater barrier and trip amplifier, with certified $MTBF_d$ figures of 1000, 2000, and 1250 years and standard deployment, the sensor sub-system $MTBF_d$ would be 128 years:

$(0.001 + 0.0033) + (0.0005 + 0.0011) + (0.0008 + 0.0011) = 0.0078/year.$

If we allocated 35% of function PFD to this sub-system, we would need to test every 2.8 years to meet the same mid SIL1 target. With increased vulnerability the sub-system $MTBF_d$ would be 53 years with a corresponding test interval of 1.17 years.

(0.001 + 0.01) + (0.0005 + 0.0033) + (0.0008 + 0.0033) = 0.0189/year.

## Conclusion

These results appear sensible; in essence they show that there would typically be no difficulty in meeting SIL1 with a single channel with standard deployment and that SIL2 would be a realistic prospect provided vulnerability was reduced or testing frequency was increased.

## References

1. Reliability Data for Safety Instrumented Systems, PDS Data Handbook 2010 Edition, SINTEF.

## 21. VALIDATING FAILURE RATES AND MANAGING END OF USEFUL LIFE

There is a requirement in IEC 61508 for *'...assessing whether the demand rates and failure rates during operation and maintenance are in accordance with assumptions made during the design of the system.'* (Part 1, Clause 6.2.12 c, see also Figure 8) This requirement propagates into IEC 61511 as *'Discrepancies between expected behaviour and actual behaviour of the SIS shall be analysed and, where necessary, modifications made such that the required safety is maintained.'* (Part 1, Clause 16.2.9) Easy to say, but how may we do this on a practicable basis in a real process plant operation? In principle it is a straightforward matter; you identify the relevant equipment populations and monitor the failure rates. In practice, the identification of the populations, and analysis of failures may be less than straightforward.

The point of the monitoring is to identify where a failure rate is higher than anticipated, perhaps because the assumed intrinsic device failure rate was optimistic or because the specifics of the equipment deployment increase the failure rate, or the equipment is entering the wear-out phase and approaching its end of life.

Any failure of equipment within a SIL rated function should be thoroughly analysed to identify the cause of the failure and the possible implications for other similar equipment items deployed on SIL rated duties. It would be unrealistic however to expect the same degree of analysis to extend to equipment not deployed on SIL rated duties, particularly the typically much larger set of equipment deployed on control and monitoring (rather than protection) duties. Much of this equipment will 'repaired-by-replacement' and detailed analysis of any failure is likely to be an unrealistic ambition. But much of the equipment deployed on control and monitoring may well be the same as that deployed on SIL rated duties. This wider population set, in providing a broader sample, will be potentially useful in identifying equipment failure rates.

From the perspective of integrity, it is the undetected dangerous failure rate that matters, and in principle it would be possible to estimate and monitor the dangerous failure count. Since the split of safe/dangerous failures in the non-SIL population is unlikely to be available we might estimate the number of dangerous failures on the basis of the total number of failures and an estimate of the Safe Failure Fraction (SFF). This estimate might be combined with any explicitly identified dangerous failures in the SIL population to identify an estimate of the total number of dangerous failures within the wider population. There are potential difficulties here however; undetected dangerous failures are only revealed by proof tests, and some proof tests may be at extended intervals, and a given element type might be subject to a range of test intervals. Note also that the classification as a safe/dangerous failure may depend on the duty e.g. high/low trip function. Since the SIF population would typically be a small subset of the broader population, there would likely be marginal advantage over a simple total failure count. If there is a population of items dedicated to SIF duties there may be some advantage in assessing the dangerous failure rate rather than the total failure rate. Generally it will be simpler to monitor a count of failures (whether safe or dangerous), with the working assumption that any significant shift in dangerous failure rate would also manifest in a higher number of safe failures.

Failures might be monitored from maintenance work order records and spares consumption as part of an annual review by the site maintenance authority. This analysis would be for all equipment types that are also deployed on SIL duties and so one of the first requirements is the compilation of a register of such equipment types.

With this register there is the difficulty of knowing how far to go in identifying the actual build and deployment of equipment items. Consider an ESD valve; do we distinguish actuator from valve? Different sizes of a given type? Different material combinations? Different duties for given build? Different environments for a given build and duty? The greater the resolution in categorising build and deployment, the smaller the populations available on which to base our analysis.

It is here proposed that initially we should identify populations as far as manufacturer and series type, together with vulnerability due to the specific nature of the deployment. Typically we would identify manufacturer and series of a flow meter, but not what size or material combination. This would typically mirror the assessments undertaken by manufacturers which are generic to a series design type. Only if a failure is subsequently found to be only relevant to a particular subset would I propose a greater resolution in categorisation of populations.

Since actuators of different types and from different vendors may be used with a given valve, it could be argued that actuators and valves should be distinguished as different elements. There are practical difficulties in this however and generic database values make no such distinction. Again, only if a failure is subsequently found to be only relevant to a particular subset of actuators would I propose to attempt to distinguish the actuator as a separate element. Seat failure to provide tight shut off (as distinct from stroking failure), will typically arise through service life and associated wear, and should be addressed by assessment of useful life expectancy rather than random hardware failure rate. Given this consideration, attempts to refine valve equipment groupings on the basis of shut off requirements probably represent an unwarranted complication.

## VALIDATING FAILURE RATES

If the duty is unexceptional, it is suggested that no distinction be made beyond manufacturer and series type. If a particular vulnerability is identified for an item due to its deployment/duty, then a special sub-grouping should be identified for that equipment series type.

In practice it might be difficult to identify whether a given failure in the more vulnerable subgroup was attributable to the particular vulnerability or whether it was a 'normal' failure within the wider population of that equipment type. It is proposed therefore that a failure in the subgroup should be recorded against both the subgroup AND the wider population superset. A failure outside the subgroup would be recorded against the wider population only.

Let us postulate a total population of model type 'Acme Z800' of 90, 8 of which are identified as being on 'difficult duty 1', and 5 on 'difficult duty 2', which increases their vulnerability. With one 'normal duty' device failure and one in each of the 'difficult duty 1' and 'difficult duty 2' groups, the record would show:

| Device | Group | Population | Failure Count |
|---|---|---|---|
| Acme Z800 | All duties | 90 | 3 |
| Acme Z800 | Difficult duty 1 | 8 | 1 |
| Acme Z800 | Difficult duty 2 | 5 | 1 |

**Table 21.1** Failure Count Categorisation

This approach will avoid an optimistic bias in the evaluation of standard duty deployment arising if we were to disregard possibly relevant failures in subgroups. It will be conservative with regard to subgroup populations, since any failures that are not associated with the duty will have more apparent significance in the smaller population.

I would not propose identification of firmware revision as a basis for sub-dividing populations for monitoring of ongoing reliability. I would only expect to consider firmware revision as part of change management, or where upon analysis a failure was found to be due to vulnerability of a particular firmware revision.

## Detecting increased failure rate

If the actual number of failures in a population differs markedly from that expected, this may be an indication of wear out or an error in the original estimate of failure rate. If the number of failures differs by more than a nominated margin we may prompt revision of the PFD/PFH calculations or consider the implications for equipment selection/replacement. I suggest that actual number of failures should be compared with the anticipated number of failures on an annual basis. If there is an apparently significant increase in failure rate from that anticipated, a representative equipment item should be inspected to see whether wear out mechanisms have begun to influence performance; if so, all similar equipment should be overhauled or replaced. If the equipment is relatively new or if there is no evidence of wear out, it may be that the original estimate of failure rate was wrong. (The anticipated number of annual failures is simply the population multiplied by the failure rate in years.)

In order to provide suitable discrimination the following threshold criteria are suggested for possible end-of-useful-life alerts based on comparing actual number of failures with the anticipated number:

AMBER alert: actual failures >= number of anticipated failures x 2, rounded, minimum 1.

RED alert: actual failures >= number of anticipated failures x3, rounded UP, minimum 2.

Unless the number of anticipated failures in a year is very low (<0.02 say?), in which case raise a RED alert with a single failure. (If the probability of a failure is so low, if it does happen we should consider why.)

If two amber alerts occur in consecutive years, or if a red alert occurs, the equipment should be inspected at the next opportunity. The point of the amber alert is that there is always the potential for a 'blip' in performance with subsequent regression to the mean.

Note that if the population is small and the reliability is high, failure count monitoring will not be useful as a means of detecting potential end of useful life; the anticipated number of failures will be so low that the failure rate (and associated PFD) could rise significantly without failures being observed.

**Identifying Useful Life**

IEC 61508 (Part 2 7.4.9.5 Note 3) says that typically equipment will have a useful life of between 8-12 years, but the provenance of this note is unclear; it might be based on the typical useful life of aluminium electrolytic capacitors which are sometimes identified as the limiting component in electronic equipment. Certainly there are many items that remain in service well beyond this period without exhibiting wear out. Vendors may provide an indication of useful life, but understandably will qualify this with the need to consider the specific duty the equipment is deployed upon. In the absence of specific information, and with industrial grade equipment operating well within its operating limits, perhaps a useful life of 12 years might be assumed (coupled with a wear out provision as detailed below) unless there are contrary indications (e.g. known severe service/environment) or arguments to substantiate a greater period.

Failure mechanisms associated with 'wear out' may involve physical wear and tear due to erosion/abrasion/creep/fatigue or from degradation of materials due to chemical mechanisms such as oxidation. In the absence of significant cycling stress or creep effects, if an equipment item has no moving parts (typical of electronic equipment), or parts that rarely move, it may be that chemical degradation mechanisms may dominate. In which case we might employ the Arrhenius equation to qualify useful life as a function of operating temperature. The Hallberg-Peck equation is commonly used to estimate an acceleration factor ($A_f$) for life testing of electronic equipment:

$$Af = \left[\frac{RH_R}{RH_U}\right]^3 e^{\left\{\frac{E_a}{K}\left[\frac{1}{T_U}-\frac{1}{T_R}\right]\right\}}$$

Where:

$RH_R$ = Relative Humidity at Reference Conditions

$RH_U$ = Relative Humidity at Conditions of Use

$T_R$ = Temperature at Reference Conditions (°K)

$T_U$ = Temperature at Conditions of Use (°K)

$E_a$ = Activation energy in electron-Volts

$K$ = Boltzmann's constant (8.623x10-5 eV/ °K)

This combines the well-known Arrhenius equation for reaction rate with a cubic factor from the ratio of relative humidity under test conditions to that in use. Reference 1 identifies an activation energy value of 0.9eV as generally recommended in the absence of specific data on failure mechanisms. Reference 2 identifies 0.6eV as 'A good general activation energy for many classes of failures…'

If useful life is assumed to be inversely proportional to reaction rate, we can use this relationship to estimate useful life in service from declared life under reference operating conditions:

$$Service\ Life = Reference\ Life \times \left[\frac{RH_R}{RH_U}\right]^3 \cdot e^{\left\{\frac{E_a}{K}\left[\frac{1}{T_U}-\frac{1}{T_R}\right]\right\}}$$

Without the Hallberg-Peck qualification for relative humidity, the above equation reduces to:

$$Service\ Life = Reference\ Life \times e^{\left\{\frac{E_a}{K}\left[\frac{1}{T_U}-\frac{1}{T_R}\right]\right\}}$$

It should be noted that these equations are relatively crude estimates (results are particularly sensitive to the value of activation energy), but having recognised that useful life may be influenced by operating conditions, they do provide a rough and ready basis for accounting for operation at conditions significantly removed from declared reference conditions. If nothing else they may prompt critical inspection earlier than might otherwise be scheduled.

Recognising the difficulty in identifying the appropriate value for activation energy without information on the specific failure mechanisms, we may approach this from a different direction. If a device has an operating range of 0-100C say and a reference temperature of 50C, with a warranty period of 1 year, we can back-calculate what the corresponding activation energy is that would cause sufficient acceleration to reduce useful life from a nominal 12 years to say 2 years (warranty period +1 year) when operating at the upper operating range limit. (The notion being that a device operating at the operational limit would be expected to survive beyond the warranty period.) By transposition of equation 3:

$$Activation\ Energy = \frac{\ln\left[\frac{Warranty\ Period + 1}{Reference\ Life}\right] \times K}{\left[\frac{1}{T_U} - \frac{1}{T_R}\right]}$$

For the figures quoted above, the corresponding activation energy would be 0.37eV. We may then use this value in calculating the actual service life. Assuming nominal 12 year life at a reference temperature of 50C, the service life at 70C would be 5.5 years.

The very large uncertainties in this approach must be acknowledged, but it does allow the estimation of useful life on a consistent rational basis.

If the equipment life is likely to be dictated by the number of operations, the commonly used B10 specification may be used. (B10 is the number of operations after which 10% of the population is expected to have failed.)

This is typically used to estimate the failure rate used for probability of failure calculations:

$$\lambda \approx \frac{0.1 \times N_{OP}}{B10}$$

Where:

$N_{OP}$ = operation/cycle frequency

$\lambda$ is the total failure rate

If a B10 figure for dangerous failures is nominated, this can be used as above to directly identify the dangerous failure rate.

The B10 specification can be used to estimate useful life:

$$Life \approx \frac{B10}{N_{OP}}$$

Note that the B10 specification cannot be sensibly used if the equipment life will be determined by degradation mechanisms other than cycling e.g. oxidation. Low demand functions will not typically see sufficient operational cycles for the B10 specification to be useful; if B10 is used as above, nonsensical failure rates and useful life periods would be identified.

Wear out will typically manifest in a progressive fashion and although a device may have reached the end of its nominal useful life, that is not to say it is immediately untenable. It would be unrealistic to attempt to model wear out on a specific device-by-device basis, but we may adopt a generic approach in recognition of the behaviour. We may postulate a power law increase in failure rate with time:

$$\lambda = \lambda_i \times \left(\frac{SL}{NUL}\right)^a$$

Where:

$\lambda$ is actual failure rate beyond nominal useful life

$\lambda_i$ is initial failure rate during useful life

a is a constant

SL is actual service life

NUL is nominal useful life

A cubic law (a=3) gives an approximate doubling of failure rate with a 25% extension in service life beyond the nominal useful life, and 8 times if service life doubles useful life. The actual value of 'a' is likely to be a matter of judgement, (unless hard data is available), but for want of anything better the above figure is suggested as a starting point (a doubling of failure rate would typically be of limited significance given other uncertainties, and a 25% extension does not appear ambitious). The typical expectation is that wear out will be progressive (no 'cliff edge') and there are many examples of equipment continuing to provide good reliability performance at 20+ years despite the declaration in the standard of typical useful life of 8-12 years. Unless an extension to useful life is substantiated, the PFD calculation should be revisited and test intervals revised as appropriate.

If the assumed failure rate is increased to the point where the anticipated number of failures (within the deployed population) is 1/year, (i.e. when the MTBF in years = population) there need be no further increase in the assumed failure rate, since failure count may then be relied upon to indicate a higher failure rate.

If there is a mix of ages in a population there is the possibility that an increased failure rate in the older subset might be masked by the younger. Clearly, it would, in principle, be possible to monitor the failure count in these subsets independently, but this may well be too complicated an approach. Alternatively, the failure rate of those safety elements that are older than their nominal useful life may be assumed to progressively increase as indicated above, with PFD recalculated correspondingly and test intervals adjusted if appropriate. A swap of younger for older elements might also be considered.

Logging of the ages of all (SIL + non-SIL) element populations is perhaps an unrealistic ambition. But it is suggested we should maintain an age database of all SIL rated elements. For each deployed SIL element, the 'vintage' and function tag should be recorded. Vintage being the 'birthday' of the element, whether by virtue of an overhaul or new installation. If end of useful life is detected or assumed end approached, this will allow identification of all similar elements of that vintage.

## Conclusion

It is possible to conceive of all sorts of possible refinements here, but we need to establish a practicable basis, not fantasise about what might be. There are many uncertainties in risk assessment and functional safety concerns and it is pointless to concern oneself with the third significant figure (or even possibly the second). I would suggest that only if an identified failure rate changes by a factor two or more would it be appropriate to consider revisiting PFD/PFH calculations, useful life assumptions, or equipment selection options. The failure rate of modern equipment is so low that random failures will be relatively infrequent unless there is a large population. In practical integrity terms, the dominant issues are likely to be systematic influences, and management and maintenance practices. Nevertheless, it is appropriate to establish a practicable basis for identifying possible end-of-useful-life or misplaced initial assumptions and hopefully the above will help in this regard.

## References

1. SSB-1: Guidelines for using plastic encapsulated microcircuits and semiconductors in military, aerospace and other rugged applications, Henry Livingstone, Diminishing Manufacturing Sources and Material Shortages (DMSMS) conference, 2000.

2. Highly accelerated stress screening for air-cooled switching power supplies Part 1: Understand stress test methodology White paper: TW0058, Ron Weglinski, Transistor Devices Inc., February 1, 2007

## 22. SHARING OF ELEMENTS BETWEEN BPCS & SIS

The sharing of elements between BPCS and SIS is often a vexed question. This can be in respect of field equipment; either sensors or final elements, or signal processing. Let us consider first the field equipment aspects, and to do this in less abstract terms, I examine here a familiar arrangement; a modulating BPCS control valve in series with a remotely operated isolation valve (ROV), both triggered to close by the same SIF. Some will immediately declare this to be a violation of the principle that BPCS and SIS provisions should be independent; this is certainly true, but the question is 'does this signify?' Independence of SIS and BPCS should be pursued wherever practicable, but the sharing of elements is not prohibited.

The actual stipulation in IEC 61511 is:

**11.2.10** *A device used by the BPCS shall not be used by the SIS where a failure of that device may result in both a demand on the SIF and a dangerous failure of the SIF, unless an analysis has been carried out to confirm that the overall risk is acceptable.*

*NOTE when a part of the SIS is also used for control purposes and a dangerous failure of the common equipment would cause a demand on the function performed by the SIS, then a new risk is introduced. The additional risk is dependent on the dangerous failure rate of the shared device because if the shared device fails, a demand will be created immediately to which the SIS may not be capable of responding. For that reason, additional analysis will be necessary in these cases to ensure that the dangerous failure rates of the shared devices are sufficiently low. Sensors and valves are examples where sharing of equipment with the BPCS is often considered.*

Before diving into 'additional analysis' of the pfd however, we should first make a simple test of the architecture requirements; does the SIL target derived from the initiating event due to the dangerous failure rate of the shared element(s) **alone** require hardware fault tolerance. If it does, the sharing of a control valve in a notional 1oo2 architecture would not be compliant, since in response to the control valve failure the architecture would only be 1oo1. The implication is that we would need a second ROV (unless we can demonstrate this would be beyond ALARP). Note that the SIL target arising from shared element dangerous failures alone may be less than that for **all** initiating events.

Typically BPCS dangerous failures, constrained at no better than 1 in $10^5$ hours, will dominate the initiating event frequency and may well be the driver for a SIL target that requires hardware fault tolerance. Note however that for this particular architecture evaluation we do not need to use the constraint on BPCS failure rate claims; in our example, if the control valve is triggered to close by the SIF acting on a solenoid valve, we need only to consider the dangerous failure rate of the shared element; the control valve. This should be significantly better than the 'BPCS' figure (or the usual constraint would be in question); this is legitimate as long as the shared element is managed as a part of the SIL rated provision.

If we are satisfied that the architecture is acceptable, the question now is; how to perform the 'additional analysis' of pfd? Some experts will urge a full quantitative risk analysis upon you but in truth the evaluation can be made on a relatively straightforward basis.

Often the approach adopted for evaluating the risk reduction achieved by a non-independent arrangement, is to use one of the following two methods:

- Assume the non-independence has negligible effect; assume a full 1oo2 architecture for pfd calculation. (Optimistic)

- Disregard the redundancy and calculate pfd on the basis of a 1oo1 architecture. (Conservative)

For a shared control valve, typically the first approach would inflate the evaluation of achieved risk reduction by a factor two or so, which might not be considered a gross distortion, particularly if the achieved RRF is well up in the target SIL band.

A more refined approach is to combine the two methods and calculate the RRF achieved with each architecture ($RRF_{1oo1}$ and $RRF_{1oo2}$) and combine the results as follows:

Take the shared element dangerous failure rate as the initiating event frequency for which the 1oo1 architecture may be claimed ($IE_{1oo1}$), and the total initiating event frequency minus the shared element dangerous failure rate as the initiating event frequency for which the 1oo2 architecture can be claimed ($IE_{1oo2}$).

The overall risk reduction factor can then be calculated as:

$$Overall\ RRF = \frac{Total\ IE\ Frequency}{Total\ Post\ SIF\ Frequency} = \frac{(IE_{1oo1} + IE_{1oo2})}{\left(\frac{IE_{1oo1}}{RRF_{1oo1}} + \frac{IE_{1oo2}}{RRF_{1oo2}}\right)}$$

This will provide a refined estimate of RRF and therefore PFD and allow assessment of compliance in that regard.

The above discussion is in relation to field equipment; sensors and final elements. What if the BPCS control processing and SIF 'logic solving' are themselves implemented in the same unit? The first piece of guidance in relation to sharing of BPCS control and logic processing is simple; don't.

It is allowed, but it brings a variety of headaches in engineering, verification/validation, maintenance, management of change etc. That said, the arrangement is often met, particularly with packaged plant items such as boilers. A particular point to note is that failure of the SIF processing would potentially also mean a failure in control processing, leading to an excursion to which the SIF could not respond. Since in these circumstances the hazard potential would be realised upon failure of the SIF, the SIF (in respect of the shared processing) is operating in 'continuous mode'. If the target is SIL2 this would immediately require redundancy under route $2_H$ or under route $1_H$ if the SFF is less than 90%.

In verifying the reliability aspects of compliance there would be two components; one associated with the continuous mode processing with a corresponding PFH, and a low demand PFD associated with BPCS **only** failure or other initiating event demands that are external to the control processing.

Hazard rate (H) will be:

$$H = \frac{[PFH_{processing} + PFD.D_{ext}]}{RRF_{post\ SIF}}$$

Where $D_{ext}$ is the demand rate due to events external to the processing unit, and $RRF_{post\ SIF}$ represents the risk reduction due to any protection layers that would be invoked after a SIF failure (e.g. mechanical relief would typically be invoked after a high pressure trip layer).

If we assign a Tolerable Frequency (TF) to the hazard potential, by transposition we can identify, for the low demand component:

$$REQUIRED\ PFD = \frac{[TF.RRF_{post\ SIF}] - PFH_{processing}}{D_{ext}}$$

And for the continuous mode component:

$$REQUIRED\ PFH_{processing} = [TF.RRF_{post\ SIF}] - PFD.D_{ext}$$

Note that the two components are interlinked (and consistent units must be employed). The SIL target for the function should be identified as the higher of the SILs corresponding with the requirements for each of the two components. The PFH for the processing will be immediately fixed by the hardware selection for the duty and would then fix the required PFD for the SIF from the low demand mode component.

## 23. TRIP SETTING NOMINATION & PROCESS SAFETY TIME

Any Safety Requirements Specification (SRS) worthy of the name will identify the trip setting and the process safety time (PST). This latter is defined in the IEC 61508 standard to be the 'period of time between a failure, that has the potential to give rise to a hazardous event, occurring in the EUC or EUC control system and the time by which action has to be completed in the EUC to prevent the hazardous event occurring'. (EUC stands for 'Equipment Under Control'.) This is not quite right however. Consider an example of a cooling water failure; this will lead to a temperature rise, but the initial rise before the trip point is reached does not constitute part of the PST. The PST is the time between the trip setting being reached and the time by which the action must be complete if the hazard is to be avoided. The process safety time is then useful as a component of the SRS in that it identifies the maximum acceptable total execution time of the Safety Instrumented Function (SIF).

Note that if the protection function is a permissive interlock (i.e. preventing an operation), there will not be a process safety time. If the function is for mitigation (e.g. fire detection), rather than prevention, typically the response time will not be an issue. If the function is a trip derived from detection of a binary status condition; drive on/off, valve open/closed, flame/no flame etc., then PST is determined directly by considerations of the process and plant design. If the trip is derived from a continuous process variable however e.g. pressure, temperature, level, then the PST becomes a function of the trip point nominated. The farther from the hazard point the trip setting is, the greater the PST. (Note that this is also true when a switch is used on a continuous process variable – the trip point is implicit in the switch setting or level location.)

Often the trip setting will be nominated as a judgement based on past practice and experience rather than any rigorous evaluation. Typically there will be a handsome level of conservatism in the specification of the true process limit e.g. equipment pressure or temperature rating, and a lack of conservatism in the nomination of a trip setting would not be potentially hazardous. For many applications, conservatism in the specification of a trip setting would not be a critical concern, but for some there may be profound implications for process performance and availability. It may be that the closer the process may approach a constraint the better the process yield or efficiency. This is one reason for improving process control; it may allow a set point closer to the trip setting through reduced process variability. A critical examination of the trip setting nomination may identify opportunities to revise the trip setting itself and allow operation with a reduced margin to the process limit.

The PST will also often also be nominated on the basis of established practice and judgement rather than any formal evaluation, but although it may not be recognised, implicit in a specification of PST for a continuous variable trip is the approach speed of the variable to the true constraint, since:

$$PST = \frac{(PV\ at\ Latest\ Acceptable\ Completion\ Time - Trip\ Setting)}{Variable\ Approach\ Speed}$$

The actual trip point may differ from the trip setting because of uncertainty in the measurement comparison between the trip setting and the process variable:

$$Actual\ Trip\ Point = Trip\ Setting + Trip\ Point\ Error$$

The influence of trip point error is often so small that it may be disregarded, but this cannot always be assumed to be the case.

Explicit identification of the speed of approach to the constraint will allow a refined trip setting specification to be determined on the basis of the protection speed of response. Progressive throttling during shut-off, for example, may well mean that the approach speed will reduce once the protection is invoked, but typically a worst case linear approach speed would be used to estimate the trip margin. A full analysis of approach trajectory would have to include inertial effects and process dynamics and this degree of rigour would only be employed in exceptional circumstances.

A rule of thumb that is often adopted is to aim by design for the SIF response time (SRT) to be no more than half the PST. This is not an inviolable rule however; design considerations might mean that something longer than half is appropriate. Slavish adherence might well lead to the specification of larger actuators for instance, with unwarranted consequences for size, weight and expense. A discussion with the process engineer might well reveal that the declared PST is 'negotiable'. Even if the original PST is confirmed, a response time greater than half may be perfectly acceptable as long as there is confidence that the overall trip execution time will not grow to exceed the PST.

The appropriate trip setting for a process variable will be identified with a margin to the process limit and may be influenced by a number of considerations:

- The post trip increment in the process variable due to process lag e.g. fill line drain down inventory adding to a level, or temperature continuing to rise due to multiple order temperature lags.

# TRIP SETTING NOMINATION & PROCESS SAFETY TIME

- Uncertainty in the process variable measurement and the trip point

- Uncertainty in the specification of the process limit (e.g. bursting disc rupture)

- The amplitude of the process noise i.e. of the uncontrolled higher frequency fluctuations in the process variable; a trip point must be at least half this amplitude away from the hazard point.

The characteristics bearing on trip settings are illustrated in figure 23.1:

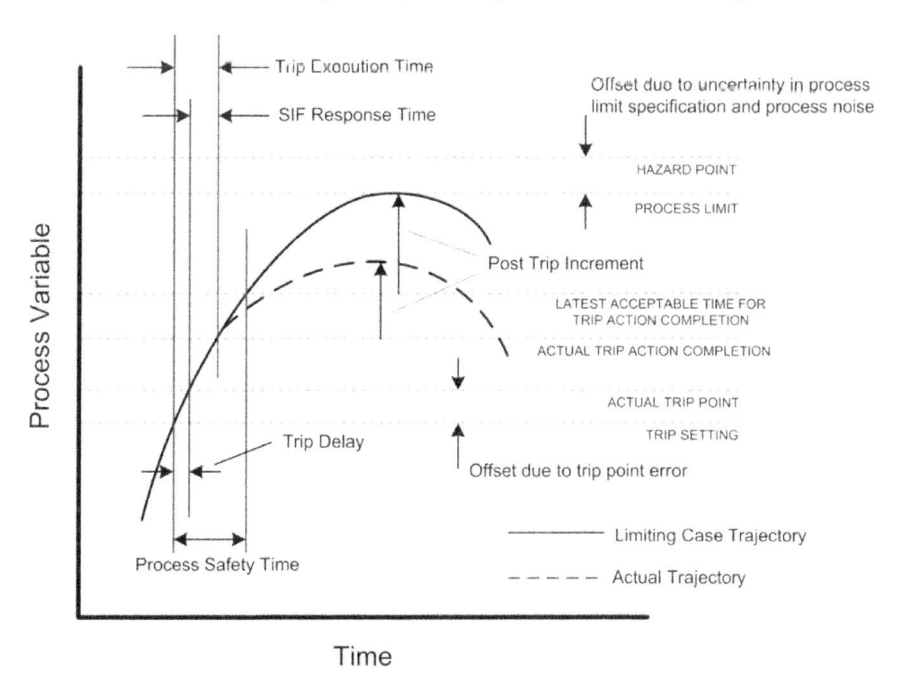

**Figure 23.1** Trip Setting Characteristics

There is potentially some confusion over what constitutes the Hazard Point. As an example, if we consider loss of containment due to overpressure and rupture of the pressure envelope as the ultimate hazard, we will likely include mechanical relief as an independent layer of protection in our SIL determination analysis. The hazard point is then the potential rupture pressure and the true process safety time is the time to this rupture pressure; essentially we would be designing the SIF as a protection layer to cater for the possibility of mechanical relief failure. Note however that in designing a SIF to support this it is possible that the pressure excursion post trip could still trigger the mechanical relief even though this was not necessary to supress the hazard. If avoidance of mechanical relief was a critical operational concern a new constraint on trip execution time would arise. If the hazard point was identified as the relief setting we would identify a shorter process 'safety' time; this would be conservative but might give rise to unwarranted difficulty in the SIF design. A critical review of the hazard specification and design options would be indicated.

The uncertainty in a trip point will be a function of the capability of the equipment; it will be determined by the accuracy and drift of the sensor system and the trip amplifier and the associated calibration interval. Typically the uncertainty of the trip amplifier will be so low relative to the process measurement uncertainty that it may be disregarded. The uncertainty may cause a trip to be delayed or advanced relative to the nominal trip setting. If the actual trip is closer to the constraint than the nominal setting, the trip will be delayed by a time corresponding with the trip point error and the speed of the process.

$$Potential\ Trip\ Delay\ =\ Trip\ Point\ Tolerance \div Variable\ Approach\ Speed$$

$$Guranteed\ Trip\ Execution\ Time\ =\ Potential\ Trip\ Delay + SIF\ Response\ Time$$

The trip point tolerance (i.e. potential trip point error) may be established from a published safety specification identifying the appropriate tolerance, or it may be calculated from equipment performance specifications. This is not necessarily a straightforward matter however, a rigorous calculation must combine accuracy and drift specifications of the system components together with the calibration interval, and include installation effects and pertinent influence quantities such as equipment operating temperature and process operating conditions; this is beyond the scope of the present text. Note that some safety specification tolerances might well be an order magnitude greater than those simply identified by the more usual performance (accuracy) specifications. The performance and calibration of instrumentation systems is often identified using a 95% confidence level corresponding with two standard deviations of a normal distribution. This implies a 1 in 40 chance of a dangerous out of tolerance value from this consideration alone. This is not consistent with SIL performance requirements. It is here suggested that a tolerance established from published specifications (and incorporating drift, installation effects and influence quantities) should therefore typically be expanded by at least a factor two. (Giving a tolerance at approximately 99.994% confidence.)

The ultimate requirement is that the SIF response time should not exceed the PST minus the potential delay due to trip point error.

$$Maximum\ Allowable\ SIF\ Response\ Time = PST - Potential\ Delay$$

A more refined rule-of-thumb as a design target would be to say that that the SIF response time should be no more than 50% of the maximum allowable.

Without this refinement it is conceivable that a design could appear to be satisfactory with an SRT of less than 50% of the PST, but potentially unsafe in that the trip point tolerance could mean an additional potential delay of more than the remaining PST. The 50% design rule makes allowance for increased SIF response times in the installed system. There is nothing substantiating the 50% figure however, it represents a judgement of what is a prudent allowance. If the design is found to breach the above rule of-thumb (or is otherwise considered to be possibly insufficiently robust in terms of the timings), the options are:

- Engineer a reduced SIF response time

- Engineer a reduced trip point tolerance

- Consider whether the values for the process limit and/or approach speed may be revised

- Change the trip setting to increase the margin from the process limit

- Use more rigour in the analysis to demonstrate that the **guaranteed** trip execution time (i.e. that for which the declared failure rate used in the probability of failure on demand calculation is valid) is less than the PST.

## Conclusion

The widely employed rule-of-thumb that SIF response time should be less than 50% of process safety time is potentially deficient in that it does not take account of a number of subtleties in the characteristics of trips relating to continuous process variables; in particular the uncertainty in trip points. A more discriminating rule-of-thumb is to stipulate that the SIF response time should be less than 50% of the value that would otherwise cause the SIF execution time to match the PST. Breaches of this rule are not necessarily hazardous however and a more rigorous analysis of the system provisions may well demonstrate that values in excess of 50% are perfectly acceptable.

## 24. COMPARISON ALARMS

Comparison of measurements may be used to allow early detection of measurement system faults, with the potential to reduce production losses and enhance protection function integrity. Automatic comparison and alarming can be implemented in a simple manner using conventional function block capability as typically available within distributed control systems (DCS). An outline of the functioning of such a system is described, together with details of how credit may be taken for deployment of such an alarm when performing protection function probability of failure on demand (PFD) calculations.

**Measured Variable Comparison (MVC) Alarm Approach**
Where more than one measurement of the same process variable is available, typically one associated with control and one with a trip function, it is possible to compare the measurements to establish a degree of confidence that the sensor sub system associated with the trip function has not been compromised. A simple deviation alarm might be used to alarm when an excessive discrepancy arises between the measurement signals, but this arrangement is susceptible to momentary discrepancies that might arise from noise within the measurement signals. The alarm threshold may be increased to counter this, but this is not an entirely satisfactory approach. One alternative that has been proposed is an alarm based on a statistical Cumulative Sum calculation. Some vendors offer such algorithms, but these are sophisticated tools

that may require specialised resources to deploy and may well incur additional expense. A simpler approach with which to overcome the nuisance alarm problem is a sustained deviation alarm, in which the alarm threshold must be exceeded for a specified period before the alarm is triggered. (A simple delay timer function may be used.) A generic block diagram is given in figure 24.1. (Implementation details may differ from system to system, and the polarity of the logic may change.) Given the delay timer function, there is no requirement for dead band in the alarm block itself.

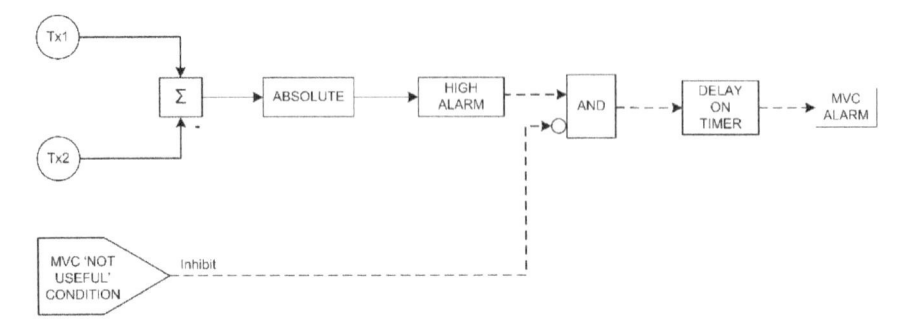

**Figure 24.1** Block Diagram

The threshold may be set at the maximum difference that would be expected when operating with two healthy measurement systems when the process is in the steady state; this then provides for differences arising from inherent accuracy limitations. To allow the use of a single alarm, the absolute difference should be used; this may be by use of a function or calculation that selects the absolute value of the difference signal. Note that it is proposed that the threshold should correspond with the maximum discrepancy to be expected from healthy systems in the steady state, rather than the maximum acceptable discrepancy (without compromising the trip function). Use of the healthy maximum will help with early identification of deterioration in the measurement systems. In the absence of operational experience, the value may be conservatively nominated from published accuracy specifications for the measurement systems. The alarm will then be triggered when the difference signal exceeds the threshold by more than the prevailing noise band (in the difference signal) for more than the specified duration. (Noise that carries the signal below the threshold would cause the delay timer to reset.) A noise band of ±1% would mean that the alarm would operate when the measurement signal is maintained above the alarm setting +1%.

The delay time needs to be set sufficiently long to avoid spurious nuisance alarms due to transient discrepancies that will arise from healthy measurement systems. These may arise from noise effects or from differences in time response in the measurement systems. The delay should be several times longer than the period of any dominant noise component and longer than the expected duration of transient discrepancies due to a difference in measurement lag times. (And certainly therefore longer than any difference in measurement dead time lags.)

Measurements selected for comparison will typically be of similar type and location such that there will not normally be a large difference in the responsiveness of the measurements. The size of any transient discrepancy would depend on the differences in the measurement time lags, and the process excursion rate and duration. The appropriate delay may perhaps be most readily identified empirically by observing the process behaviour. (The time spent above the healthy steady state discrepancy threshold during plant excursions.) Delay time is not critical provided it is long enough to suppress nuisance alarms. Although long delay times could be used to suppress nuisance alarms, the delay time should not be extended arbitrarily since this would delay the alarming of a genuine alarm condition and the prompt to investigate and rectify faults. Delay times are typically expected to be of the order of minutes rather than hours.

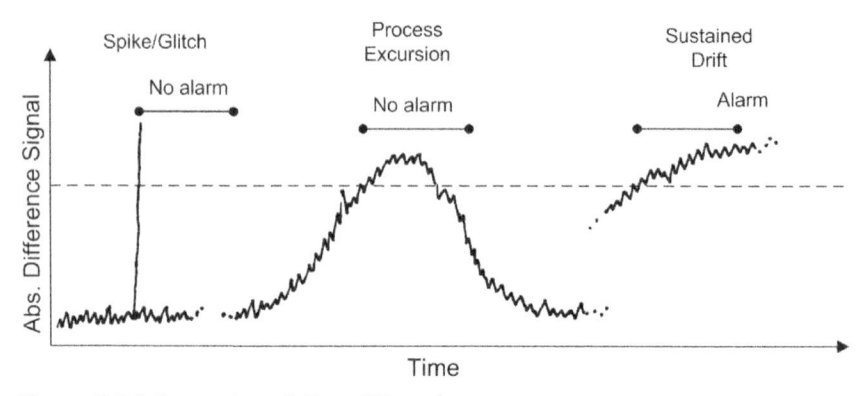

**Figure 24.2** Examples of Signal Trends

Potential nuisance alarms due to these transient discrepancies could also be suppressed by raising the alarm threshold, but this would also compromise the ability of the comparison to detect deterioration in the measurement systems and is not the preferred approach. Differences in measurement system scaling should be accommodated in the difference calculation, but for effective comparison the measurement ranges should overlap for all operating regimes that may persist and where a demand might be placed on the protection. (Otherwise operation could persist at a condition where the comparison is effectively disabled.) If there are anticipated operating circumstances where an out of range condition would generate a nuisance alarm, an automatic inhibit on detection of this circumstance may be configured e.g. valve position or control set point. If these circumstances could persist AND could give rise to a real demand on the protection, then an inhibit of the alarm is not an acceptable approach; the measurement provisions should be revised to remove the out of range condition. Detection of the out of range condition itself should **not** be used to inhibit the alarm, since the out of range condition could be due to a measurement failure rather than operating point changes. Trip points should be biased by the discrepancy threshold to allow for the possibility that a discrepancy of that magnitude could persist and not be alarmed. To reduce common mode potential it would be preferable to route the measurement signals via different control system input modules.

## Probability of Failure on Demand Contribution

A measurement comparison alarm is effectively a diagnostic provision that will detect some of the dangerous failures in the sensor subsystem of a trip function that would otherwise have been unrevealed. It will not however detect common mode failures that influence both measurement systems equally.

We can estimate the effective diagnostic coverage provided by an MVC alarm as follows:

Diagnostic Coverage (DC) = $\lambda_{DD}/\lambda_D$

Where $\lambda_{DD}$ is the dangerous detected failure rate

And $\lambda_D$ is the dangerous failure rate

Given common mode factor β, the dangerous detected failure rate may be established from the **non**-common mode dangerous failure rate x the probability of a successful alarm response.

$\lambda_{DD} = (1-\beta).\lambda_d.(1-P_{ca})$

Where $P_{ca}$ is the probability of failure on demand of the MVC alarm and the required operator response.

Giving

DC = $(1-\beta).(1-P_{ca})$

With $P_{ca}$ typically assigned a value of 0.1, we may identify the following representative values:

| β | DC | $\lambda_{DU}$ |
|---|----|------|
| 5% | 85% | 0.145 $\lambda_D$ |
| 10% | 81% | 0.190 $\lambda_D$ |
| 20% | 72% | 0.280 $\lambda_D$ |

**Table 24.1** DC and $\lambda_{DU}$ Values as a function of β

Substitution of these values in sensor subsystem PFD calculations will then allow proper determination of the system PFD with the nominated test intervals.

Note that we have identified here the DC due to the MVC provision which will include the process connections if these are included in the nomination of $\lambda_D$ and $\beta$; it is suggested that this DC figure should be used in lieu of any figure identified for a trip transmitter/device alone. (It is to be expected that the comparison with an independent measurement will provide more coverage than possible from intelligent monitoring provisions in a single device.) If DC, Safe Failure Fraction (SFF) and total failure rate ($\lambda_T$) are declared for a device, $\lambda_D$ **for the device** may be calculated as:

$$\lambda_D = [(1-SFF_{OLD})/(1-DC_{OLD})] \cdot \lambda_T$$

This figure should be increased to account for process connection dangerous failures if appropriate. (See Chapter 20 for a discussion of such considerations.)

If required, given $DC_{NEW}$, $SFF_{NEW}$ may be calculated as:

$$SFF_{NEW} = 1 - (1-DC_{NEW}) \cdot (1-SFF_{OLD})/(1-DC_{OLD})$$

If a device had an $SFF_{OLD}$ of 50% with no diagnostics, ($DC_{OLD} = 0$), then an MVC alarm with common mode factor of 10% would give an $SFF_{NEW}$ of 90%

## Conclusion

The approach outlined in this chapter may offer a relatively straightforward tool with which to enhance a trip function, requiring only configuration of a DCS or equivalent. Some care is needed to ensure that the nuisance alarm rate is suitably low, so that the alarm remains credible and will prompt investigation and repair as appropriate, but that said, the tuning of the alarm is relatively straightforward and robust. Claims made for the alarm will be constrained by the reliability of the operator response and the nature of the system in which the alarm is configured.

Assuming a common mode factor of 10%, and a typical total system PFD contribution from a sensor sub-system of 30-50%, we can anticipate overall system PFD will reduce to approximately 75% to 60% of its value without a comparison alarm, assuming test intervals remain the same. Conversely, the deployment of a comparison alarm may allow an extension of function test intervals (by a factor 1.33' to 1.66') to a more practicable frequency. If the original test frequency for other subsystems was maintained, the sensor subsystem test interval could be increased by a factor of 5.

The gain is modest, but still potentially useful in some circumstances. The alarm will also help identify degradation that might otherwise lead to a spurious trip.

## 25. PROOF TESTING

This chapter explores some of the possibilities in a pragmatic approach to proof testing of process plant safety instrumented systems, with particular regard to the functional safety system standards. It is preferable that a proof test should reflect real operating conditions; in this respect, the ideal approach is to drive the process variable to a point where it triggers the safety system action, but without placing a real demand on the protection, i.e. with no risk of a hazard should the protection fail during the test. In practice end-to-end testing with 100% coverage is usually difficult or impossible to implement on a practicable basis on real plant.

The following observations on approaches to proof testing are by no means exhaustive, but are intended to highlight the considerations that will allow users to identify the specific requirements for their own installations. Recognition of these same considerations should also allow more effective design of an installation to facilitate proof testing.

'New build' systems should be specifically engineered to accommodate proof testing requirements. Existing installations should be critically reviewed to identify where upgrades are appropriate.

**Underpinning Philosophy**
The overall requirement is to establish a proof testing regime that will identify all dangerous failures within an appropriate timescale to ensure

the safety function Probability of Failure on Demand (PFD) requirement is met.

It should be recognised that the indiscriminate pursuit of the proof testing 'ideal' may actually degrade safety – it may compromise system integrity and/or divert resources which might be better employed in securing a more significant safety return elsewhere. The law of diminishing returns applies. It should be remembered that there are no absolute guarantees of safety – from the moment a proof test is completed there is a possibility and associated probability that the function has suffered a dangerous failure. Our concern is with the prudent and responsible management of risk, **not** its elimination (which is impossible). With testing, as with other safety system concerns, the higher the SIL, the greater the degree of rigour required.

The purpose of proof testing is to drive the probability of failure on demand, due to random hardware failures, to a sufficiently low level consistent with the nominated SIL. Its purpose is **not** to validate the design. (Design validation is an earlier part of the safety lifecycle). The ongoing integrity of the design should be maintained by appropriate 'management of change' provisions. At some point, testing to re-validate the design may be appropriate, but this is not the purpose of routine periodic proof testing.

There may well be a trade-off between coverage factor (i.e. the proportion of dangerous failures that will be detected by a test) and the amount of disturbance to the installation. (See Chapter 37 for a discussion of coverage factor nomination). A disturbance to an installation to allow enhanced coverage may risk inadvertently compromising the protection function, or may expose personnel to other hazards e.g. with the breaking of process connections. The risk associated with the disturbance may mean that the marginal increase in coverage is not warranted or should be undertaken at extended intervals. Note that coverage here relates to the potential dangerous failure mechanisms rather than the physical extent of the system. As an example, if possible failure mechanisms for an emergency shutdown valve are identified (together with relative contributions) as; seizure in the open position (20%), solenoid valve failure (20%) or poor shut off (60%), a partial stroke test would identify the first two categories of failure, but not the third. Coverage achieved with the partial stroke test would be 40%. (Consideration would also have to be given to all other failure mechanisms relevant to the function.)

If a partial proof test is implemented more frequently in recognition of the partial coverage, the requirement to perform a full test to provide 100% coverage can be at extended intervals. If a full 100% proof test would be required at period T to achieve the required probability of failure on demand, a partial test at twice that frequency (period T/2), would mean that a full 100% (Maintenance Interval) check would only be required at an extended interval if the same PFD is to be maintained. We may establish the ratio of the partial proof test period and the full maintenance period as a function of the partial proof test coverage and the frequency of the partial proof test. This relationship is summarised in the table below for a partial test at twice the required frequency of a full 100% maintenance check and test. The derivation is given at the end of this chapter.

| Coverage Factor (%) | Ratio of Partial Proof/Maintenance Interval |
|---|---|
| 60 | 3.5 |
| 70 | 4.3 |
| 80 | 6.6 |
| 85 | 7.7 |
| 90 | 11 |
| 95 | 21 |
| 98 | 51 |

**Table 25.1** Test interval ratio as a function of coverage factor

So if a proof test with 100% coverage would be required every year, the same PFD could be achieved with a partial test at 90% coverage every 6 months, and a full maintenance check every 11 x 0.5 = 5.5 years. The maintenance check might use removal and overhaul to the 'as new' condition in lieu of an in-situ test.

These considerations may allow difficult proof tests to be undertaken at extended intervals (typically coincident with scheduled plant shutdowns) with simpler intermediate proof tests performed at higher frequency.

These considerations should not be used to extend a maintenance or full test interval unreasonably. PFD and reliability calculations are conventionally based on random failures, but some failure mechanisms are associated with service life. The argument given above should not be used to extend test/maintenance intervals to the point where in service degradation/ageing mechanisms become a significant factor. In addition, more frequent partial testing may accelerate wear and tear and this could result in a requirement for more frequent maintenance.

PFD calculations are beyond the scope of this chapter, but should recognise the uncertainty associated with reliability data, the possible operational variation in proof test intervals, and the difficulty of identifying the failure modes of complex components.

It may be appropriate to capture information from real trip events. This may enhance coverage or avoid the need to shutdown to perform a difficult proof test. (If a real trip was demonstrated to be successful, there would be little point in shutting down to perform a scheduled proof test the following week.) Critical examination of historical trends may provide evidence, but care is needed to avoid misinterpretation: Did a flow stop because: the control valve closed, the pump stopped, the ESD valve closed or some combination of these? Did a control valve close because of a command from the DCS or operation of an ESD solenoid valve? If an effect has multiple possible causes, it may be difficult to confirm the initiation by a particular cause. Cascading trips as a plant shuts down may invoke the effects from causes other than the one of interest. The possible ambiguities make reliance on *ad hoc* feedback from operations personnel an unsatisfactory approach. An unequivocal determination is required; the basis and required records should be identified as part of a formal, documented method. This approach is only likely to be useful in exceptional circumstances.

**Test Method Principles**
Disturbance to the physical installation should be minimised to reduce the potential for the inadvertent introduction of failures. Testing should be non-invasive as far as is practicable e.g. circuit checks for presence/absence of voltage or continuity are preferred, where no disturbance to circuit wiring is required. Such tests should be cycled through presence/absence to prove correct circuit addressing, otherwise a 'successful' test could be recorded when the test probes are inadvertently applied to the wrong terminals.

Any disturbance should be failsafe wherever practicable e.g. a failsafe circuit may be isolated from supply by wire disconnection or opening of a terminal knife. Overrides should be avoided as far as practicable – brief energising of circuits to perform cycle tests may be via a hand held flying lead, rather than by modification to the circuit installation. The wholesale application of overrides to force a system to a healthy condition should be avoided. Consideration should be given to the possibility of upgrading the system design to facilitate testing and reduce the need to disturb the installation.

If override jumpers are used for any reason, they should be highly conspicuous (e.g. bright colour, with minimum wire length of say 1m) so that there is minimal risk of them being inadvertently left in place. Their application/removal should be recorded in a register.

If an element is demonstrated to be in a safe (tripped) state, it may not be necessary to see it cycled to/from the un-tripped state – if the element has in fact failed to the safe state this would be revealed when a process start is attempted. But bear in mind that a check on speed of response might be required if this is critical to the safety function.

It may not be necessary to prove the operation of a final element for each independent initiator. Sometimes initiator circuit contacts are wired in series in a final element supply circuit (rather than via a separately identifiable logic solver sub system), in which case, once the final element itself has been proven, it will normally be sufficient to prove that each contact in the supply circuit goes open circuit upon operation of the associated initiator. A break at the correct point in a final element failsafe supply circuit should be confirmed. A supply circuit might be de-energised for a number of reasons – confirmation of a de-energised final element state does not necessarily confirm operation of the protection function.

Where hazardous fluids are involved, any interference with the process connections (e.g. to perform in-situ calibration or to allow removal of a device to a workshop) is itself potentially hazardous to personnel, or may, for example, allow introduction of moisture which might ultimately compromise the measurement. There is also a residual risk of the connection not being correctly restored which is a potentially dangerous failure on some measurements e.g. high pressure trips. Under such circumstances, it may be preferable to perform a simple (but formally recorded) observational check that the measurement is 'live and credible'. I.e. that the measurement is responding and not frozen for some reason, and that the value is consistent with plant conditions and other measurements.

Continuous process measurements (as distinct from switched signals) may have their calibrations checked by comparison with other measurements of the same plant variable. Far from being a less rigorous approach, this may help identify potential failure mechanisms, such as blocked impulse lines, which would not be apparent with a workshop calibration check. Consideration should be given to possible common cause errors however (particularly the possibility of calibration/ranging errors) and the reliance placed upon the protection provision (how high the SIL). If measurements are made using different technologies there would be a corresponding increase in confidence in the comparison.

Historic trend data may also be used for cross checking of measurements. A periodic (and formally recorded) review of readings when the plant is running may reduce the need for calibration checks when performing protection system proof tests with the plant shutdown.

Note that inaccuracies at other than the trip point do not impinge on the protection function and multi-point calibrations may be of limited value. A second or third point check will be useful in highlighting any gross non-linearity that might indicate a failing device or ranging error, but a full upscale/downscale 5+ point calibration of every measurement may not be warranted.

Testing in-situ is preferable wherever practicable. If an equipment item is to be removed for overhaul purposes – testing should be conducted **after** reinstatement. In-situ calibration checks are also preferred. Removal to a workshop should be confined to critical measurements that cannot be validated in-situ, or where removal is necessary for other reasons e.g. maintenance overhaul. If there are failure mechanisms that are difficult to effectively test for, it may be that to achieve the appropriate coverage it may be necessary to perform an overhaul of an item to restore it to the 'as new' condition.

Following a calibration check of the sensor, signal injection may be used to test the input circuit to the logic solver. A direct mA injection may be preferable to the use of a hand held terminal with a 'smart' device, since the connection/disconnection of a mA injector would be a failsafe disturbance. It would seem perverse to deny use of the calibration mode of a smart device however; the critical requirement is to have positive affirmation of a return to on-line measurement mode. (Possibly by cycling power off/on to force a return to default on-line mode.)

It is preferable to use an injector that can be operated in measurand engineering units (e.g. degC, m3/hr) rather than signal levels (mA, V, %), since this avoids the potential error in conversion and would prompt investigation of any discrepancy due to a ranging error. This is one immediate advantage of using a 'smart' device in injection mode.

Trip amplifier settings should not normally be adjusted as a means of exercising a trip function – there is a danger that the correct setting may not be reinstated. The measurement signal should be exercised (usually by signal injection) to prove correct setting and operation.

Testing should be undertaken against identified performance standards, which should be identified in the Safety Requirements Specification. For example the operation of a trip should be within a specified tolerance of the trip setting. Calibration of measurement sensors should be within a specified tolerance, as should the stroking time of ESD valves. It is preferable that engineering units be used.

Note that the performance specification should relate to healthy equipment performance rather than the functional requirement. A healthy valve with a stroke time of 5 seconds may effectively perform its safety function even if the stroke time extends to 30 seconds, but such deterioration in performance may indicate incipient failure that should be investigated and corrected.

Testing should not normally be by placing a real demand on the protection – i.e. by forcing the process to the trip point – failure of the protection may result in an incident. This approach may be acceptable if appropriate provisions are made in recognition of the possibility of a protection failure. This might be as simple as, for example, having someone standby ready to report an overfill on a vessel. The extent and rigour of the provisions should reflect the consequences of the potential hazard.

Care needs to be taken to distinguish between a 'Functional Test' and a 'Proof Test'. It may be that a sub-system or equipment item has redundant channels; a functional test would confirm the operation of the sub-system or equipment item, but would **not** necessarily confirm whether all channels were healthy. A proof test should check the health of all channels.

## Specific Initiator Test Considerations

The following considerations are by no means exhaustive and must NOT be taken as definitively identifying the approaches to be adopted. Full consideration must be made of the local situation in the context of the required safety function.

## Pressure Transmitters

Suitable block and bleed valves on the instrument impulse line with test connections may facilitate an in-situ calibration check. In-situ pressure injection with a calibrated source will normally allow full initiator testing without the need for separate injection of a mA signal. (Injection of a mA signal will not test the actual sensor device.) If practicable, it is preferable to test by pressure injection at the process end of any impulse line connection. If in-situ pressure injection is impracticable, mA injection may be substituted for the initiator circuit test, with a separate calibration check made on the pressure transmitter.

## Temperature Measurements

RTDs are passive devices and a simple measurement check against other indications or a portable thermometer may be all that is required. A full workshop calibration is not likely to be warranted for most applications. Injection of resistance in lieu of the RTD can then be used to check the initiator circuit operation. A similar approach may be adopted with thermocouples and mV injection. Correct sensor open circuit (burnout) response should be confirmed by disconnection of the transmitter input.

## Flow Measurements

Unless there is some reason to expect a significant shift in primary device flow coefficient or k-factor (unusual in meters with no moving parts in the absence of potential for corrosion/abrasion/deposition) there may not be a need to perform a wet calibration at every proof test. Primary signal injection ('Dry' calibration) to prove correct signal conditioning may be sufficient. Visual inspection to confirm absence of physical degradation might be employed. Full wet calibration might be employed at extended intervals.

## Level Switches

Tuning fork type devices typically have a good diagnostic provision and an electronic test function may be available. If suitably configured for safe failure, the requirement to remove the switch from the process and dip the sensor may be obviated or only undertaken at extended intervals.

Similar considerations may apply to capacitance type sensors. Simple application of a finger to the sensor terminals within the head may provide sufficient capacitance shift to prove the switching functionality, and might form the basis of a useful intermediate (partial coverage) test.

## Level Transmitters

These may be most simply checked by comparison with other measurements, sight glass readings or dips. On some vessels, a portable ultrasonic detector may be used externally to detect level.

Time of flight measurements may be removed for calibration against a distance reference, but this may be of questionable value given the intelligent diagnostic provisions of such devices, and the susceptibility of the reading to process conditions and vessel configuration and internal furniture.

Hydrostatic level measurements may be treated as pressure measurements.

## Specific Logic Solver Considerations

These are typically the most reliable sub-system, and testing requirements are likely to be less onerous. Logic solving is a digital two-state operation and so not as susceptible to drift and the requirement for calibration checks. Providing good change management and access controls are in place there should be no reason to expect the functional design to be compromised. A full function check is likely to require the wholesale application of overrides and a significant fail-to-danger disturbance to the installation. This may not be warranted for some installations, or may perhaps be considered only at extended intervals.

Some solid state systems allow function testing through pulse propagation. PLC coding could be copied and tested on an off line system.

All initiator inputs should be cycled to prove their operation and correct addressing: check for presence/absence of signal/continuity at the logic solver input (if a PLC, in 'monitor' mode you may monitor the status of the relevant logic element). All outputs should be observed in the safe state. Cycling of outputs may be impracticable because of the need to apply overrides to inhibit initiators, but if the safe output state is proved (typically open circuit or supply absence) there may be little remaining prospect of a dangerous random failure and coverage factor will be correspondingly high. (With the initiators already proved, and the output demonstrated to be in a safe condition, it would need a failure within the LS that spuriously drives an output to a safe condition when input/output states change from those prevailing under test conditions – a relatively obscure possibility).

**Specific Final Element Considerations**

**Control Valve – Non Tight Shut Off**
Control valve failures are typically revealed in normal operation. Even failure of the closure member or associated linkage would normally be revealed (unlike a normally passive on/off valve). The testing requirement is on the output circuit from the logic solver.

With the process shutdown, the trip active, and with air supply confirmed as on, the valve position may be confirmed as tripped. There remains a need to confirm solenoid valve operation if the valve may have moved to the trip position for other reasons e.g. a control system command.

Options:

a) Break in supply circuit confirmed. (Solenoid spool operation untested).

b) With the control command to the control valve set to maintain fail position (if necessary), energise the solenoid with a flying lead – check solenoid operation by sound/feel. (Vent path untested).

c) With the solenoid circuit de-energised (failsafe) confirm the control valve will not operate when commanded – needs positive confirmation that operation command is active and effective – if the valve fails to respond then this would prove successful solenoid operation. Note that if the test fails, the valve will move – beware of a possible hazard. Might be implemented by isolation of the solenoid power supply at start up, with de-isolation upon instruction from operations when the valve operation is required.

d) With the valve in the healthy position, isolate the solenoid supply and confirm the valve stroke.

## On/Off Valve

If the valve is exercised as part of routine operations its failure would be revealed, and the provisions identified for a control valve would apply. If however the valve is dedicated to the protection function, a failure of the closure member or associated linkage would not necessarily be revealed. (Depending on construction and failsafe action). This is a relatively obscure failure mechanism and perhaps relates more to rotary valves than to sliding stem valves. If considered to be a concern where relatively high reliance is placed upon the valve, consideration should be given to confirmation of proper operation of the closure member itself. This might be achieved by observation of flow/pressure indications, in-situ endoscopic examination, or removal from the line. If speed of operation is critical, but susceptible to deterioration, then the valve stroke should be observed and the time to closure compared with a performance standard. Otherwise testing considerations are as per the control valve.

## Tight Shut Off Valves

Where tight shut off is critical to the protection function, it will be necessary to prove the shut off capability by observation of flow/pressure indications or by direct pressure testing. (This is often referred to as leak-off testing.) Do not rely on normal plant flow/pressure indications unless it can be shown that the test will not be invalidated by the status of other plant items; other valves, pump/compressor status etc. It may well be necessary to remove the valve to a workshop to allow an effective pressure test to be made.

### Drives
The status of a drive may be confirmed at shutdown and the protection system contact in the stop circuit may be tested for continuity to confirm it is open circuit.

### Operational Testing of Final Elements
If a final element such as a shut off valve or a contactor is exercised frequently as part of normal operation and its dangerous failure would be revealed by the operational demands, there would be no point in performing independent proof tests of this element. The PFD may be calculated on the basis of the frequency of the operational demands. There would be a requirement to independently proof test any equipment that is **not** exercised as part of the operational demand e.g. a separate ESD solenoid valve (SOV) in an air supply to an actuator that is normally controlled by a separate SOV operated by a control system.

If failure of a final element would itself create a demand on the same final element, there is likely to be a requirement for a redundant final element arrangement or other revision to the system design.

### Administration
Proof test procedures should explicitly detail the steps to be performed; simple high level statements such as 'Test loop ref XXX' should not be used.

Proof test records should distinguish between 'passed first time' and 'failed but now fixed'. There is otherwise the possibility that problems are being rectified to gain a pass result without the problem being logged/reported to the maintenance authority. The proof test record should require and facilitate recording of any rectification work or adjustment.

Proof test records should clearly describe any fault found and state if the fault resulted in a dangerous or safe failure condition.

Consideration should be given to rotation of personnel and/or independent audit of a sample of proof tests to help identify any systematic failures in procedure execution.

Proof test procedures should include for a record of physical inspection (as distinct from functional checks) to identify any significant degradation of the installation.

Completed proof tests should be traceable to the party performing the test. Proof test records should require positive affirmation of restoration of any broken connections and process de-isolations. Accountability for these actions is important in promoting discipline in test procedures.

Personnel must be competent for the tasks they are charged with.

## Derivation of Maintenance/Partial Test Interval Relationship.

Note this derivation is for a 1oo1 element or system.

If we take:

$$PFD = \frac{1}{2}\lambda T$$

For a proof test interval T, dangerous failure rate λ, with 100% coverage.

With partial coverage C and Maintenance Test Interval $T_M$

$$PFD = C.\frac{1}{2}\lambda T + (1-C).\frac{1}{2}\lambda T_M$$

Now, with $T_M$ = nT, and seeking equivalence in PFD between 100% coverage at interval T, and partial coverage C at more frequent intervals T/m, and remaining coverage at interval $T_M$:

n being the ratio of partial and maintenance test intervals. m being the factor by which the test interval is reduced (or frequency increased)

$$\frac{1}{2}\lambda T = C.\frac{1}{2}\lambda \frac{T}{m} + (1-C).\frac{1}{2}\lambda nT$$

Which yields:

$$n = \frac{(1-C/m)}{(1-C)}$$

The table within the body of the chapter is constructed for m = 2, giving n= 5.5 if coverage is 90%. So with partial proof test interval at half of 100% coverage interval, the required 'maintenance' interval would be at 5.5 x 2 = 11 times the partial test interval.

## 26. REGULATORY OVERVIEW

Many find a visit from the regulatory authority inspector to be a daunting prospect. But let us try and establish some perspective here. Like any large organisation, the HSE or the EA (or their equivalents in jurisdictions beyond England & Wales) have a spectrum of talent from the extremely able to the very much less so. Their inspectors are not demigods; they are idiosyncratic and fallible mortals. I have seen thoroughly professional inspectors, highly expert and experienced, who are prepared to take a pragmatic view of affairs. I have also seen inspectors who take an immediately hostile and pedantic stance, apparently with the idea of intimidating the inspected, and thereby establishing ascendency over them. I have seen others who, uncertain of their ground, will bluster and make 'definitive' pronouncements that are in truth nothing of the sort. I have even seen one with a sense of humour. (Just joking!) These organisations have their fair share of the egotistical, the insecure, the neurotic, the psychotic (just joking again!) None of this should come as a surprise; these foibles (and others) are to be found throughout our profession (and others). And, of course, a professional, competent, diligent, sane inspector can, like the rest of us, still get it wrong.

The encounter need not (by default should not) be adversarial; the inspector can be your friend and can help inform and guide your activity; in this respect it may be useful to engage with him sooner rather than later. There should be a presumption of good faith from both parties and the default position should be one of mutual respect.

Naturally, you should extend every professional courtesy to the inspector, as indeed he should to you. This should be the case whenever professional parties meet; whatever the niceties of conferred authority, if you wish to claim the title 'professional' you are bound by codes of professional conduct.

By virtue of his warrant the inspector has powers which include the authority to:

- enter your premises at any reasonable time
- carry out investigations and examinations
- take measurements, photographs or samples
- require an area or machine to be left undisturbed
- seize, render harmless or destroy dangerous items
- require people to give accurate information or provide statements
- inspect and/or copy any relevant documents

But these would typically only be invoked if there was an incident or want of cooperation. More usually his attendance will be by appointment as part of a routine. Typically, if some shortcomings are identified, an inspector will take informal action with written or verbal advice. Only if there are more serious failures in provisions or timely action is he likely to formally issue an 'Improvement Notice' requiring you to take action in a specified time. If there is a more immediate and significant hazard, he may issue a formal 'Prohibition Notice' requiring you to cease a specified activity. The ultimate sanction is prosecution.

He cannot however exercise these powers arbitrarily. If initial correspondence cannot resolve matters with your inspector and you believe there is an unacknowledged error or fault you may complain to his manager. If formal notices have been served you have the right of appeal. An improvement notice is suspended pending the appeal, but a prohibition notice is not.)

Remember the inspector is a public servant, paid for by your taxes and the fees he levies. He has the 'luxury' of not having to manufacture anything; he does not have to juggle operational concerns that may conflict with safety issues. (Closing a plant will make it perfectly safe, but there is a general recognition that there would typically be a net loss for society).

This is of course as it should be. The inspector must be allowed to focus on safety. The implication is that you should not necessarily expect an immediate 'meeting of minds'. The inspector will have a distinct perspective on your affairs, but you, not the inspector, are the expert on your process operation; if you believe the inspector's view to be inappropriate you should stand ready to challenge his position, and deploy suitably rational and robust arguments in support of your own view. Be aware that a less able inspector may incline to simply reiterating the formal line rather than exercising professional judgement about what is practicable. Remember that it is very easy for the inspector to spend your money! And he is unlikely to tell you that you are going over the top in your provisions.

When I say 'challenge' I do not mean seek confrontation; rather it would typically be a question of offering alternative views or possible solutions. It should be a matter of informed debate between professionals. But if you believe unwarranted burdens are being dictated you should be ready to protest. You have a right to expect a courteous and considered inspection and one would hope for good, even friendly relations, but remember that the inspector may need to maintain a certain professional distance. You may well need to investigate referenced standards and codes of practice yourself to identify just how pertinent their provisions are in the context of your operation. There may well be more than one way of meeting your obligations. I have known inspectors point to guidance which is not well suited to a duty holder's circumstances; the cited guidance or models might be useful in informing your deliberations, but should not necessarily be considered definitive and cannot necessarily be relied upon to provide the optimal approach.

It is not necessarily a question of having everything in a position to pass muster at first inspection. There may well be shortcomings in your provisions, but as long as the inspector can see that you are addressing these in a responsible and timely manner he is likely to be satisfied and content to monitor your progress. Essentially the inspector will be looking for a considered, systematic, responsible approach informed by established good practice (including pertinent standards). If he can satisfy himself that you are on the right path, albeit not yet at your destination, he is likely to smile rather than frown on your endeavours.

Genuine misapprehensions are one thing, but attempts to deny shortcomings or obfuscate are quite another and run the risk of antagonising the inspector. Misapprehensions (from either party) should be readily acknowledged, not compounded by attempts at justification with a subsequent entrenching of positions. Attempts to deceive are, of course, unacceptable and contrary to professional codes of conduct.

I do not claim the considerations I have identified here as being solely due to observation of inspectors, they are in part compiled from the wider observation of professional engineers (or even the superset of human beings I have encountered) and much of what I have written will apply to any dialogue between professional parties.

## 27. FUNCTIONAL SAFETY MANAGEMENT PLANNING

*I keep six honest serving-men (They taught me all I knew); Their names are What and Why and When, And How and Where and Who...* (Rudyard Kipling)

Functional Safety Management Planning is a critical aspect of project execution but is typically not handled well. There is a tendency for Functional Safety Management Plans (FSMP) to adopt default statements of intent and to end up with chunks of text transcribed from the standards. They can quickly become comprehensive statements about the virtue of functional safety equivalents of motherhood and apple pie; nothing to object to of course, but of no real help.

Difficulties often arise because of confusion about the role of the plan. The standards identify what should be covered in your plan, but this is typically a niche aspect of the wider project plan and the usual expectations associated with this can confuse the preparation of the more narrowly focussed FSMP.

So what should a FSMP consist of?

In essence it needs just three serving men (but they must be honest!); What-How-Who:

What (in terms of functional safety) is to be done, on what basis (how), and by whom.

- What activities are to be undertaken in support of the life-cycle phases and what deliverables are required.
- How are these activities to be undertaken; which procedures, protocols, standards and tools are to be employed.
- Who is to undertake the work and who is to be responsible for its effective execution.

Not why, when, or where (they can have the day off):

The 'why' will often be self-evident or traceable to the standards, and even if you see good reason to record the rationale for pursuing a particular approach, that need not appear in the plan itself. It may be captured in some other document but it is perhaps better not to pad out the FSMP and possibly make the useful stuff harder to see.

'When' is not in the domain of the FSMP but rather that of the project execution plan. I would not expect to meet a Gantt chart in the FSMP. The dependencies of the life-cycle phases are identified in the life-cycle model, but this does NOT translate to a critical path for functional safety planning. The standard presents the lifecycle as a so-called 'waterfall' model, with each phase cascading, in serial fashion, to the next downstream phase. Back in the real world we should recognise that project execution involves iterative loops and parallel processing paths. The 'whens' of functional safety are necessarily integral to the 'whens' of the wider project plan, and if you attempt to separate them you will likely find yourself obliged to duplicate them, with corresponding scope for discrepancies to develop.

'Where' is really a matter of logistics and typically will be implicit in the 'how' of the FSMP and can be detailed if appropriate in the project execution planning.

Another potential pitfall is to try and make the FSMP all encompassing. Rather than a monolithic edifice, a modular approach is likely to be more manageable, with the FSMP establishing a framework that references independently maintained documents.

The honesty of our 'serving men' is important; we should avoid kidding one another (or ourselves) about the plan. If we don't yet know, we should say so. It may be that some aspects of the plan can only be identified at a later point. The plan may well need to evolve and should not be considered as 'cast in stone' at the outset. That said however, we should be sceptical of too much talk of its being a 'living document' if this is to be invoked as justification for procrastination. The details may be coloured in later, but the outline sketch should be in place from the start.

The following prompts may be useful in identifying the scope of your FSMP, but this should not be considered a definitive model:

**Project Scope**: Project boundaries.

**Project Authorities:** Who carries what responsibility?

**Related Engineering Contracts:** Who and scope (services rather than products).

**Competence Management**: Protocol/Basis for management of competence and where assessments held.

**Hazard & Risk Assessment:** What approaches, what basis?

**SIL Determination:** How? Tolerable risk nomination.

**Design & Engineering:** Policy, standards, approach.

**Supplier Assessment:** Who/How?

**Action Tracking:** How?

**Tool Evaluation:** What, how?

**Verification Approach:** Protocol and tools for check and approval e.g. documents, equipment, systems, application programs, configurations,

FAT/SAT, installation, commissioning.

**Functional Safety Assessments:** Protocol for FSA; what stages, who, what basis, what approach?

**Audits:** Protocol for audit of compliance with project protocols/standards.

**Training:** What/who/how?

**Modification/Change Control:** How?

**Security:** What provisions to be made for security from unauthorised interference.

**O&M Requirements:** Provisions for ongoing maintenance and performance monitoring.

## 28. AGGREGATE RISK & RISK PROFILES

In evaluating aggregate risk, use may be made of one or more 'hypothetical persons'; a model person with a nominal work pattern that is chosen to be representative of the risk profile that individuals might experience. The profile is typically chosen to be representative of the 'most exposed worker'; the idea being that everyone else who metaphorically stands behind this hypothetical worker will be less exposed, so if we manage the risk for the most exposed worker, everyone else will be OK. This works as long as the risk profile of each less exposed person is a subset of the most exposed. Two shift workers who work the same shift pattern on the same site would potentially have different risk profiles if they worked on different plants with different hazards and never swap roles. If we manage the risk for the shift worker in plant A, this may be of no help to the shift worker on plant B.

Working out the aggregate risk for our hypothetical person can be a significant challenge. They may be exposed to different hazards at different times as they move about a site or as plant conditions change. If the hazards are sparsely distributed, the worker may only encounter one at a time; if not, he may encounter more than one simultaneously.

One simple manoeuvre that is used, is to lower the tolerable risk target for individual SIFs to reflect the possibility of being exposed to multiple hazards – a factor ten is commonly used. Although you may have a corporate risk target for individual fatality of say $10^{-4}$/year, you would use $10^{-5}$/year for the SIL determination of any individual SIF.

This is usually very conservative, and would bring about a corresponding inflation in SIL targets. Workers are not typically exposed to 10 separate fatal hazard scenarios simultaneously; they may meet 10 as they move about the site, but not all at the same time.

Some judgement may be made about the appropriate factor to employ in recognition of the potential for simultaneous exposure.

A more sophisticated approach is to estimate the aggregate risk for the risk profile of our hypothetical person. To do this we need to map the site with the fatal hazard zones for each fatal hazard. If these zones do not overlap we have a sparse hazard distribution and we do not need to consider simultaneous exposure. Where there is a significant degree of overlapping we may proceed as follows:

We can label each distinct sector of our map:

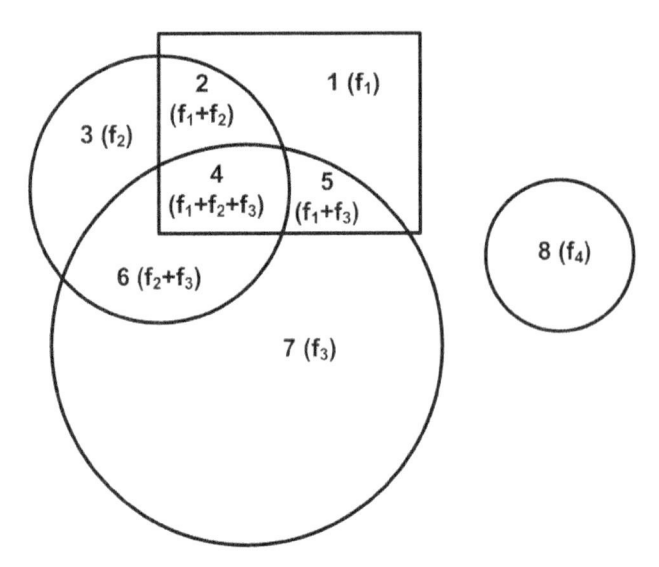

**Figure 28.1** Hazard Map Schematic
(Here we have four hazards that will be realised with frequencies $f_1$-$f_4$).

We can then estimate the percentage of time spent in each sector by our hypothetical worker ($Occ_j$).

The fatal hazard frequency can be calculated for each sector as:

$$F_{sj} = \sum_{i=1}^{n}(f_i \times Occ_j)$$

Where:

$F_{sj}$ is the total fatal hazard frequency arising in sector j

$f_i$ is the frequency of the fatal hazard event arising in the sector from hazard i (and is therefore the target tolerable frequency for the hazard event)

$Occ_j$ is the occupancy; the percentage of time the hypothetical worker spends within sector j.

Note that occupancy percentage should include a factor for the fraction of time spent on-site, typically 20%. So all the occupancies, including that in the 'no-hazard' sector (where f=0), should add up to 20%. E.g. if our hypothetical worker spends a quarter of their on-site time in a given hazard sector, the occupancy in that sector should be declared as 25% x 20% = 5% (of the person's total time). Occupancy may be based on the proportion of time within an area associated with known activities e.g. loading/off-loading operations. In the absence of activity-time-weighting, time on-site may be simply allocated in proportion to estimates of sector area as a fraction of the overall operational area of the site. (Square counting from a map superimposed on a grid might be used.)

The aggregate individual risk (AIR) is then the sum of all the individual sector exposures:

$$AIR = \sum_{j=1}^{n} F_{sj}$$

(With j=1-8 in our example.) This can be compared with the corporate target for individual risk. The individual hazard event target frequencies ($f_i$; $f_1$ - $f_4$ in our example) can be adjusted until the AIR meets the target. This adjustment would be an iterative exercise at the discretion of the user. It might be found that increasing the frequency target of some hazards and reducing that of others might allow a more practicable SIF engineering solution; the risk reduction requirements would change in proportion to the changes in the target frequency. The attractions of a spreadsheet approach will be recognised.

Note that since occupancy is already factored into the target frequencies, it should **not** be included as a conditional modifier claimed in SIL determination. It is the adjusted target frequency ($f_i$) that should be used in LOPA or QRA as the tolerable frequency for each hazard scenario. Risk graphs or matrices do not lend themselves to this approach because they do not make use of an explicit target frequency. The target frequency is however implicit in the calibration of the risk graph/matrix; if this is identified it is possible to 'work backwards' using the demand frequency etc. to work out what the corresponding SIL target should be. Note that exposure/occupancy should be assumed as 100%.

Like everything else in functional safety, we should avoid over-elaboration here. The point is to demonstrate that we have thought about these issues sensibly; the mapping should support suitably informed judgement. We look for nominal values only, we do not aim for rigorous mathematical precision. Opportunities to consolidate and simplify should be embraced with enthusiasm.

## 29. EVALUATION OF COMPOUND SIF

Sometimes we may meet a scenario that has multiple initiating events (all contributing to the likelihood of a given hazardous event), but which have **different** SIF invoked as defences. Under these circumstances, these different Sub-SIF act in combination to reduce the risk. The combination can be construed as a compound 'Super-SIF'.

Consider the hypothetical example of a direct fired heater or furnace in which the hazard scenario is re-ignition of unburnt fuel following flame failure. The flame failure may be due to a number of initiating events (IE). Let us assume the following proportions:

a) 30% caused by low airflow
b) 10% caused by undetected failure of the gas control valve
c) 30% caused by low gas pressure
d) 30% caused by high gas pressure

We can do a LOPA study and identify that a risk reduction requirement corresponding with a SIL3 function, say, is identified to reduce the risk to the tolerable risk for the scenario (TR)

To defend against this hazard potential, we may postulate the following sensor sub-systems:

# FUNCTIONAL SAFETY IN PRACTICE

1. Flame failure detection (2oo3)
2. Low air flow (1oo2)
3. Low gas pressure (1oo1)
4. High gas pressure (1oo1)

In each case the defence is to trip the fuel valves (1oo2), so in that respect the logic solver and final element sub-systems are common to all four sensor sub-systems. We can conceive of a block diagram that looks like this:

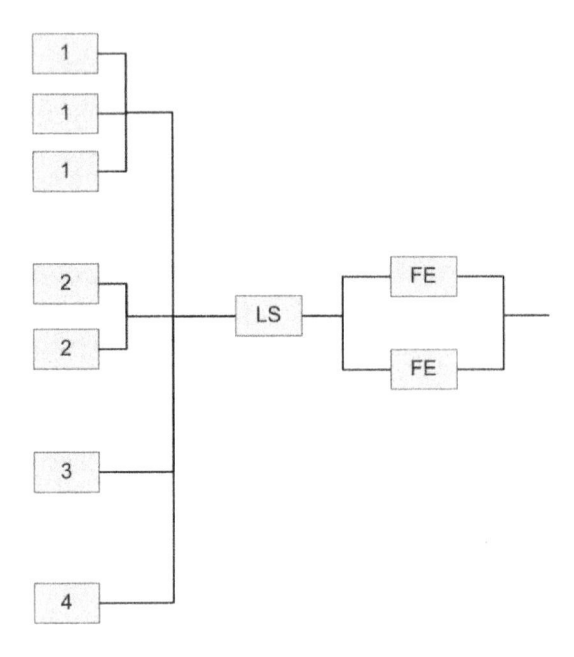

**Figure 29.1** Compound SIF Block Diagram

But this is not an accurate representation for evaluation purposes because not all the sensor groups defend against all the initiating events. Sensor 1 will defend against all four initiating events, but sensor 2 will only defend against IE a), sensor 3 against IE c) and sensor 4 against IE d).

The question now arises; how are we to verify compliance of the design of such a compound SIF? There are a variety of difficulties here; sensors 1 and 2 offer Hardware Fault Tolerance (HFT)=1, sensors 3 and 4 have HFT=0. The systematic capability (SC) of the sensors might all be SC=2 say. It is apparent there are a number of non-compliances with the SIL3 target here.

The 'obvious' solution would be to make all sensor subsystems at least 1oo2 (HFT=1) and only use elements with SC=3, but that might well be impracticable and it may place an unwarranted burden on the design.

There may be some room for manoeuvre here by distributing the overall tolerable hazard event frequency target between the individual Sub-SIFs in proportions that do NOT correspond with their initiating event frequency (IEF) contributions. (If the proportions did so correspond, the ratio of IEF to TF for each Sub-SIF would remain the same as for the overall Super-SIF, and the SIL target for each Sub-SIF would remain the same as for the Super-SIF.) The key requirement is that in aggregate, the Sub-SIF continue to meet the overall tolerable hazard event frequency.

Imagine two Sub-SIF each handling 50% of the overall IEF (assume 1/yr for simplicity) and an overall SIL target RRF of 1200 to reach the TEF of 1/1200 per year (0.000833/yr). We can postulate two solutions, one requiring two SIL3 Sub-SIF, and one requiring 1 x SIL3 and 1 x SIL2 for example:

| SOLUTION | SIF | IEF (/yr) | RRF | TEF (/yr) | SIL |
|---|---|---|---|---|---|
| A | Sub-SIF1 | 0.5 | 1200 | 0.000417 | 3 |
|   | Sub-SIF2 | 0.5 | 1200 | 0.000417 | 3 |
|   |   |   |   | 0.000833 | **TOTAL TEF** |
| B | Sub-SIF1 | 0.5 | 800 | 0.000625 | 2 |
|   | Sub-SIF2 | 0.5 | 2400 | 0.000208 | 3 |
|   |   |   |   | 0.000833 | **TOTAL TEF** |

**Table 29.1** Alternative Solutions for Same Total TEF

The SIL2 requirement may be a much more practicable proposition for the Sub-SIF 1 sensors, and the higher RRF burden placed on Sub-SIF 2 may be no embarrassment. The SIL3 requirement on the logic solver and final element would of course remain.

In our example, the four Sub-SIFs each use the same logic solver and final elements, but the compound sensor subsystem used sensors as follows:

Sub-SIF$_a$: 1 & 2

Sub-SIF$_b$: 1

Sub-SIF$_c$: 1 & 3

Sub-SIF$_d$: 1 & 4

We can now evaluate the Sub-SIF reliability block diagrams for compliance (in terms of PFD, HFT and SC) with the individual Sub-SIF SIL targets. Sub-SIF$_d$ for example looks like this:

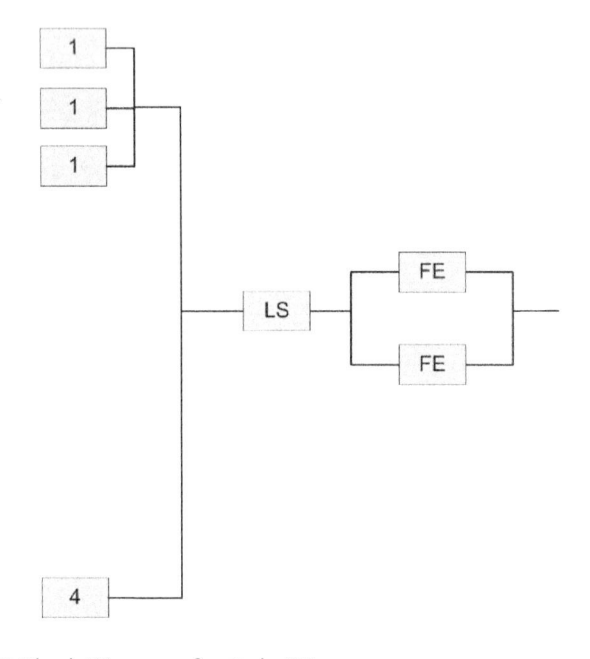

**Figure 29.2** Block Diagram for Sub-SIF$_d$

In summary then:

- The risk reduction target depends only on the hazardous event and the frequency of occurrence without the SIF, and is not affected by the design of the SIF.

- If we consider a full LOPA for the hazard scenario (all causes and related IPL etc.), this will identify the RRF (& SIL target) for the 'Super-SIF'. The Super-SIF happens to have a 'compound' sensor subsystem, but that is by-the-by; the SIF design is not defined at this point, just what it has to deliver in terms of risk reduction.
- Since the LS and FE are common to all the Sub-SIF they have HFT and SC requirements corresponding with the Super-SIF SIL target (SIL 3 in our example).
- The Sub-SIF may have their individual risk reduction factor targets adjusted, as long as in aggregate the overall target is still met; some may be reduced - provided others are increased. (Iterative adjustment of targets may be needed to converge on an appropriate solution - a 'goal seek' approach with a spreadsheet may be useful.) This adjustment may then allow the SIL target for a Sub-SIF to be changed with potentially less onerous HFT and SC requirements for the associated Sub-SIF sensor subsystem. (Note also that for those Sub-SIF that use diverse sensor types (a, c, d) it may be acceptable to use an SC further reduced by 1 from the Sub-SIF SIL since a 'sufficiently independent' combination would allow the effective SC to be uplifted by +1.)
- Proof test intervals for the LS and FE would have to be the shortest identified in the PFD calcs for each Sub-SIF.

Consider now compounded final elements. A typical circumstance would be a blending vessel fed by multiple lines, but only by one at a time. To protect against an overfill we can conceive a SIF whereby a high level sensor commands closure of a valve in each feed line. Imagine 10 identical feed lines, all presenting the same initiating events and the same contribution to overall risk. The SIF would be engineered to command close all 10 valves, but the final element subsystem should not be analysed as 10oo10, but as 1oo1. Only the valve in the line causing the potential overfill needs to close to supress the hazard. The LOPA will identify the SIL target for the overall SIF (and therefore for the sensor subsystem and logic solver), but a mini-LOPA with one-tenth of the overall tolerable risk with a 1oo1 final element subsystem would reveal a

SIL target of 1 level less for the final element.

If the protection was analysed as 10 individual 'mini-SIF' each with a SIL 1 target, the implication is that the common sensor subsystem should be 'promoted' to SIL 2. If the protection was analysed as 1 Super-SIF with a target of SIL 2, the implication would be that the individual FE subsystems could be 'demoted' to SIL1. This consideration could have a profound impact on project specifications. Some care is needed in nominating the scenario to be analysed using LOPA; this nomination should be made with a full understanding of the implications for SIF specifications.

We may also have the converse, with 10 tanks, say, all being fed (one at a time) via a manifold from a single feed line that includes the final element, but here the typical approach would be to have a single LOPA for an overfill hazard arising from any tank. The overall PFD calculation should be based on a 1oo1 sensor subsystem (NOT 10oo10 or 1oo10!)

For the purposes of the overall PFD calculation therefore, the individual tanks may be aggregated and modelled as one representative tank subsystem. The 10 sensor subsystems must still support the target SIL in terms of HFT and SC however, since although each tank contributes only a tenth of the risk they also have a more challenging target of a tenth of the tolerable risk.

It is conceivable, if there are a sufficient number of mini-SIF, that the requirement on individual subsystems could fall below SIL 1; I suggest it would be prudent to maintain a minimum SIL 1 requirement however in order to preserve an appropriate designation for management purposes.

Note that these compound SIFs should not be confused with multiple but independent SIFs all employing the same final element for example. If the multiple SIFs are truly independent i.e. defending against unrelated hazard scenarios, then the analyses can stand alone. The only additional consideration is that any equipment employed in multiple SIFs must support the highest SIL target from amongst those SIFs.

## 30. LEADING INDICATORS AND FSA4

According to the HSE's guide, *Developing Process Safety Indicators* [1], 'Lagging indicators show whether the outcome has actually been achieved.' 'Leading indicators highlight whether the risk control systems in place to deliver the outcome are operating as designed.' If we were discussing process regulatory control systems, we would speak of feedforward (leading) as distinct from feedback (lagging). The lead/lag timing aspect is with respect to the desired outcome from any given risk control provision – it is not necessarily associated with the ultimate hazard being protected against. So the distinction is between the 'number of system failures' (lagging) and e.g. 'number of tests/inspections/maintenance tasks performed to schedule' (leading). Data from proof tests undertaken to detect dangerous unrevealed failures (defects) in Safety Instrumented Functions (SIF) therefore constitute *lagging* indicators for the functional safety risk control provision. (In practical terms the lead/lag distinction is not that useful, the important thing is to find whatever performance measures may be effective in supporting management of process safety.)

An indicator such as 'percentage of tasks completed to schedule' would constitute a *leading* indicator for functional safety provisions, but may only weakly correlate with actual SIF performance in terms of whether they are 'operating as designed'. (They would perhaps be a better *lagging* indicator of management performance.) Possible tasks that might be monitored on this basis include:

- Proof testing
- Inspections
- Functional safety assessments
- Audits

In terms of ongoing management of SIF during the operating and maintenance phase of the lifecycle, and the corresponding Functional Safety Assessment at stage 4 (FSA4), it would be useful to have some leading indicators that are more directly related to SIF operation and that provide a means of validating the assumptions made at design.

These indicators need to be relatively simple if there is to be confidence in their longer term robustness in the face of shifts in personnel, management systems or management focus. To meet this requirement, the following candidates for monitoring are here proposed:

- Trip demands
- Pre-trip alarms
- Spurious trips
- Overrides/defeats

A simple periodic event count can be employed and compared with expectations. The counts would typically be relatively simple to identify using a historian application and can be readily analysed without recourse to the task scheduling/maintenance management system. (Which would typically be used for those indicators based on work order/task completion.) If experience deviates significantly from expectations, prompts for investigation may be raised.

The challenge then is to identify significant deviations, since there is always the possibility of short term fluctuations from an expected longer term average. This is where the Poisson probability distribution can be exploited. (The Poisson distribution expresses the probability of a given number of discrete events occurring in a fixed interval.) The cumulative

Poisson probability with which a given count may be equalled or exceeded can be established (as a function of the average rate and the event count). If the count exceeds that expected and has a relatively low probability of being randomly experienced, then we can use this as a flag to prompt investigation.

If the expectation is for an average count of 1/year, (a trip demand say) there is no surprise if we should experience a count of 1 in a year. A count of 2 in a given year is less likely, but would be no particular surprise. A count of 3 is starting to look surprising; a count of 4 would be troubling. For an average rate of 1/year the corresponding cumulative Poisson probabilities are approximately:

| Count | Probability |
|---|---|
| 1 | 63% |
| 2 | 26% |
| 3 | 8% |
| 4 | 2% |
| 5 | 0.3% |

**Table 30.1** Cumulative Poisson Probabilities for period of 1 year with rate of 1/year

If we considered the aggregate count across 2 years the corresponding figures would be:

| Count | Probability |
|---|---|
| 1 | 86% |
| 2 | 59% |
| 3 | 32% |
| 4 | 14% |
| 5 | 5% |
| 6 | 2% |

**Table 30.2** Cumulative Poisson Probabilities for period of 2 years with rate of 1/year

So here is the key point; although 3 events in 1 year might not be troubling, 6 events across 2 years would be. (These figures would of course change if the rate differed from 1/year.) The use of the cumulative Poisson probability allows a simple but discriminating analysis.

Corresponding figures may be established for a 3-year horizon. It is here proposed that monitoring across 1, 2 and 3 year horizons would be appropriate, together with warning flags if actual counts are at less than 5% probability. These flags would then correspond with 95% confidence that the count actually experienced points to a non-random deviation or error in the expected average rate. A further alert threshold at less than 1% probability would correspond with 99% confidence. (Monitoring across longer horizons would in theory provide even greater discrimination; the significance of a marginal elevation in count rate, if sustained over a longer period, would potentially be revealed, but for practical purposes associated with process plant safety the judgement here is that 3 years is appropriate.)

Expected average rates may be established from SIF design calculations (spurious trip rate) and SIL determination studies (demand rate and pre-alarm rate). By definition, for low demand SIFs the demand rate will be less than 1/year and the expected rate will normally be identified as part of the SIL determination exercise.

If the risk reduction factor expectation for a pre-alarm is a factor 10, then the frequency for pre-alarm may be assigned at ten times the demand rate. Note that this calculation is on the risk reduction factor *expectation*, rather than the risk reduction factor *claim*. The claim would typically be conservative and the expectation might well be higher, which would therefore yield a higher alarm frequency estimate.

Spurious trips would be identified as trip events where the process variable did NOT breach the trip threshold. The spurious trip rate expectation would typically be identified from safe failures in the SIF PFD verification calculations.

Override frequency expectation may be identified from anticipated maintenance activity, but for some SIFs a more appropriate measure may be aggregate override duration; a count of actual hours within the year may be compared with that expected.

These indicators are 'per SIF' and as such are primarily of use to those engineers directly charged with maintaining the functional safety provisions. As a higher level overview metric that can be deployed to higher management, a simple pie chart breakdown of the SIF population in terms of leading indicator measure probabilities e.g. >5% (green), 5-1% (amber) and <1% (red), might be used. Year-on-year trends may also be useful.

As with failure rate estimation and validation, significant uncertainties will remain; particularly in the nomination of average rates. [Some statisticians might object that some of the underlying assumptions in the Poisson distribution, in particular that of 'independent events' (that the occurrence of one event does not affect the probability of a second) are violated, but the distribution is used as a model – exact fidelity is not to be expected. Indeed, it is departure from the theoretical model that invites our investigation.]

It is incumbent upon users to make some evaluation of performance in support of FSA4 and the above is offered as a relatively straightforward, robust and practicable basis on which to proceed. If investigations do not reveal an explanation for a count that is significantly higher than expected, the implication is that the average rate is underestimated; this should prompt re-evaluation of the basis on which it was originally nominated to see whether there are corresponding implications for risk assessments, protection/control system design, or plant operations. Note that these SIF leading indicators may be effective lagging indicators for earlier defence layers.

Pre-trip alarm count is likely to be the most useful indicator in that we would expect this to be typically at least an order magnitude greater than demand rate count or spurious trip rate count, but these latter measures may well highlight issues that might otherwise be overlooked. In the hurly-burly of daily plant operation, a trip every other year may not be seen as that significant; it is only when compared with an expectation of say 1 in 20+ years that the significance is apparent. Note however that even a single count within a year would have a probability of less than 1% if the expectation was 0.01/year, which might be typical of the demand rate for many SIFs.

Regardless of the findings, the mere act of periodically looking at the data will bring the SIFs back into focus and introduce a healthy discipline to process safety management.

**References**

1. HSG254: Developing Process Safety Indicators - A step-by-step guide for chemical and major hazard industries, HSE Books, 2006.

## 31. MITIGATION SYSTEMS

Risk is understood to have two dimensions; likelihood and consequence. Those defence provisions that reduce the likelihood of a hazardous event are categorised as 'prevention'; those that act to reduce the consequence as 'mitigation'. Mitigation provisions include fire and gas detection, fire suppression, and manual initiation of emergency shutdowns. These are still 'safety instrumented systems' however and it is understandable that people may be prompted to attempt SIL determination and invoke the full safety lifecycle. The particular difficulty in SIL determination is that mitigation functions are called upon **after** an incident has occurred; the consequences of a system failure on demand is an escalation of the incident, rather than the incident itself. If a fire detection system should fail on demand, the consequence is not a fire - you already had a fire (which must be there if a demand is to be placed upon the detection), but rather a failure to respond to the fire in as timely a manner as would be the case if the detection system worked.

In SIL determination terms our concern would be with the **incremental** hazard arising from a failure of the mitigation system; the additional consequence arising from delayed intervention. You can immediately see the uncertainties and difficulties here. Similarly, manual shutdown initiation provisions are provided as a good practice provision to 'expect the unexpected', but you can see the difficulty in identifying the incremental hazard arising. Attempts are sometimes made to use SIL determination techniques such as Layer of Protection Analysis (LOPA) for this, but that is a bit like opening a can with a screwdriver; you'll get a result, but it won't be pretty!

The other difficulty typically met with in such mitigation systems is the 'integrity' with which leaks behave. It is all very well specifying a SIL3 detection system, but this implies we need SIL3 capable leaks or fires i.e. for the incidents to present themselves to our detectors at least 999 times out of 1000! With all the uncertainties in the source, rate/flux, elevation, direction, air movement, dilution, shielding etc. this is typically an unrealistic expectation. Similarly, with emergency shutdown initiation; there can be no expectation that an operator will identify the need and respond at least 999 times out of 1000.

So how are we to proceed? Clearly we need to engineer these systems with a degree of integrity and in accordance with good practice, but in terms of overall risk reduction expectations it is questionable whether anything better than a factor 10 is realistic. So even a SIL1 claim starts to look ambitious.

Nevertheless, given the safety role of mitigative systems, it is appropriate to employ many of the usual safety lifecycle disciplines. It may be appropriate to specify a performance target (PFD) for the system but without the usual expectation that this will translate into a corresponding risk reduction factor. So we might have a system performance specification of PFD ≤ 0.1 say, in support of the system design and implementation, without claiming that the mitigation offers a corresponding risk reduction factor >10.

The achieved risk reduction would depend also on:

a) the effective detector coverage - the probability that the detector subsystem will 'see' the hazard. The term 'coverage' suggests this is a question of the area under surveillance but there are many factors that may impinge on detection success as outlined above. It is a moot point whether this should be described as a probability since it is not the probability of a given release being detected, it is rather the fraction of the statistical population of release events that would be detected.

b) the achieved consequence reduction if the mitigative system were successful. This may be difficult to assess, (small fire v big fire? Injury v fatality? Big bang v little bang?), but in simple terms, if successful mitigation typically meant 1 fatality instead of 10 say, this would represent a consequence reduction of factor 10. If successful mitigation meant the hazard was eliminated the consequence reduction would be infinite.

If we model risk (R) as a simple linear function of consequence (C) and frequency (F), so:

$$R = C \times F$$

Then if we employ a mitigation system that offers a consequence reduction factor of RRF$_C$, and has a Probability of Mitigation Failure (PMF), then the full consequence will be realised with a frequency of F x PMF, and the reduced consequence (C/RRF$_C$) with a frequency of F x (1-PMF). We can then identify the mitigated risk R$_M$ as the combination of these terms:

$$R_M = (C.F.PMF) + \left[\frac{C}{RRF_C} . F. (1 - PMF)\right]$$

We can identify the overall risk reduction due to the mitigation system (RRF$_M$) as:

$$RRF_M = \frac{R}{R_M}$$

So

$$RRF_M = \frac{C \times F}{(C.F.PMF) + \left[\frac{C}{RRF_C} . F. (1 - PMF)\right]}$$

Which reduces to:

$$RRF_M = \frac{1}{PMF + \left[\frac{(1-PMF)}{RRF_C}\right]}$$

The Probability of Mitigation Failure has two components; the system PFD and the effective detector coverage (EDC) provided by the sensor subsystem. (An EDC of 90% would mean a probability of not 'seeing' the hazard of 0.1):

$$PMF = PFD + (1 - EDC)$$

If we postulate an effective detector coverage (EDC) of 90% and consequence reduction factor (RRF$_C$) of 10, we can see the impact of improved PFD on the overall mitigation risk reduction. The results are also shown for perfect mitigation (RRF$_C$ = ∞), where there is no consequence if the mitigation is successful. Risk reduction then asymptotically approaches a limit of 10, since performance is ultimately constrained by the detector coverage.

| PFD | RRF$_C$=10 RRF$_M$ | RRF$_C$=∞ RRF$_M$ |
|---|---|---|
| 0.1 | 3.57 | 5 |
| 0.01 | 5.03 | 9.1 |
| 0.001 | 5.24 | 9.9 |
| 0.0001 | 5.26 | 10 |

**Table 31.1** Overall Risk Reduction as a Function of System PFD

The table below shows the factor by which overall risk reduction improves when upgrading from PFD=0.1 to PFD=0.01 for a range of EDC and RRF$_C$:

| EDC | | | | | |
|---|---|---|---|---|---|
| 99% | 1.69 | 3.99 | 5.29 | 5.47 | 5.5 |
| 95% | 1.52 | 2.28 | 2.48 | 2.5 | 2.5 |
| 90% | 1.4 | 1.75 | 1.81 | 1.81 | 1.81 |
| | 10 | 100 | 1,000 | 10,000 | ∞ |

$RRF_c$

**Table 31.2** Factor Change in Overall Risk Reduction for System PFD Change from 0.1 to 0.01

In examining these figures, it becomes clear that specifying a system performance of anything beyond SIL1 would not typically add much overall risk reduction. (This has not prevented routine calls for SIL3 fire & gas systems!)

I have expounded the discussion in terms of fire and gas detection systems, but the arguments also hold for the likes of manual ESD initiation where 'effective detector coverage' is replaced by 'effective operator coverage', i.e. of the statistical population of events that could be mitigated by a manual ESD, what fraction would the operators be expected to respond to.

As a management manoeuvre it may be appropriate to designate the systems as SIL1, with a nominal PFD target of 0.1 say, and to invoke the corresponding functional safety management provisions. This needs some care however; there is the potential for confusion in that someone may insist on compliance with SIL provisions even when some relaxation may be appropriate to avoid unwarranted difficulties with the engineering. Alternatively, the systems may be designated as 'low integrity' safety instrumented functions and managed as outlined in Chapter 16.

Because of the concern with possible spurious operation it is usual for F&G systems to be 'energise-to-trip' rather than the fail safe 'de-energise to trip' that is usual for protection systems. This militates against high safe failure fraction which makes compliance using route $1_H$ more difficult. Route $2_H$ may sidestep this particular difficulty, but 'all else being equal', the dangerous failure rate of an energise to trip system will be higher than that of a de-energise to trip one. (It must necessarily include power failure.) Diagnostic features such as line failure detection may be employed to boost integrity.

For an examination of how mitigated consequences may be accommodated in a LOPA used for SIL determination of a **prevention** SIF, see Chapter 41.

## 32. MODIFICATION, DECOMMISSIONING AND FSA5

Truth is, I don't have much to say here, but without this chapter these aspects of the lifecycle might be considered conspicuous by their absence! And size is no measure of significance...

SIF modifications must of course be subject to proper change control, with all that implies for quality assurance, document revision and control, and project management. Notice that modifications beyond a SIF itself may nevertheless have an impact on the SIF. Modifications to plant design or operation may change the risk reduction requirements or the demand rate placed upon a SIF, and any change in the performance of any other protection layer claimed in a safety case may have implications for any associated SIF. There may be implications for maintenance and testing provisions. It may be appropriate to revisit the SIL determination for SIFs that might be affected.

Within the SIFs themselves, anything other than a 'like-for-like' equipment swap should be considered a candidate for some form of change control. Some form of proof test is also likely to be required to verify proper operation after any change, including like-for-like swaps. A partial proof test may be acceptable and possibly a more practicable option; the critical thing being that any possible impact from the change should be tested.

Change control evaluation should include a question along the lines of 'does the change have implications for protection provisions in terms of design, operation or maintenance?'.

The safety life cycle model in IEC61511 identifies the possibility of a Functional Safety Assessment Stage 5; *'After modification and prior to decommissioning of a SIS.'* But it is a moot point whether there is really any such thing as FSA5 since if the modification has any impact on functional safety the mandatory return to an earlier phase will invoke an FSA stage 3 as a minimum. I suggest FSA5 is really FSA3 in disguise:

*IEC61511 Part 1, 17.2.3; Prior to carrying out any modification to a SIS (including the application program) an analysis shall be carried out to determine the impact on functional safety as a result of the proposed modification. When the analysis shows that the proposed modification could impact safety then there shall be a return to the first phase of the SIS safety life-cycle affected by the modification.*

Note that FSA5 is identified as being **after** modification; therefore it cannot constitute the 'Prior...analysis...to determine the impact on functional safety'.

It may seem odd that decommissioning is included as part of the life cycle other than as an end point; if we are shutting the associated process down why should it matter? The critical point is that although the SIFs that are ending their life may no longer have any role in managing risk, their decommissioning may impact on what is left behind. If the end-of-life SIFs are a subset of a wider SIF population within a SIS that is to continue to operate, we must avoid compromising the SIFs that are to remain. This is not just a question of **what** is modified, but also **how**; an injudicious approach might lead to compromised availability of active SIFs through, for example, the inappropriate use of overrides or increased susceptibility to degradation. Decommissioning must be executed in a properly considered (and documented) manner. A gung-ho, slash and burn approach will not do.

## 33. THE SUSPENDED LOAD PROCESS SAFETY MODEL

The Swiss cheese model is often employed in representing process safety. The individual slices represent different defences or barriers, and the holes represent the potential for a barrier to fail. If the holes should align, all barriers will fail and the hazard consequence will be realised. Weaker barriers have bigger and/or more holes than stronger barriers. It usefully captures the idea of layers of protection, and of these being invoked in the order corresponding with the layer sequencing, but it relies upon the abstract notion of 'dynamic' holes that vary in size and location, and in illustration it requires a perspective drawing. It is an appealing illustration that immediately conveys the primary concern of multiple concurrent failures; it is perhaps less good at representing the integrated nature of process safety.

Barriers or defences that consist of multiple elements are sometimes thought of as chains because the elements must all work together if the barrier is to be effective. This is not really an accurate analogy however. 'A chain is only as strong as its weakest link', because each link carries an identical load. This is not true of process safety protection 'chains'. In terms of protection, the 'strength' of a chain 'link-element' relates to the probability of its failing (or being failed) when needed. If the strength of a link is increased, (the probability of failure reduced) the strength of the entire chain is enhanced, since the strength of a protection chain corresponds with the aggregate probability of failure of all the link-elements. The 'all work together' concept is potentially useful for our purposes in modelling protection, but the 'weakest link' notion is so strongly associated with chains that this militates against their adoption.

As an alternative, we might adopt a model of a 'suspended load' which might be considered as more complete in representing the idea of an integrated system. In this model, (which can be illustrated without employing perspective), process safety is represented as an arch carrying a suspended load that represents process hazard. The arch represents inherent safety – those design provisions that mean there is low danger level even if the active systems should fail; a load-hazard above the arch cannot be realised as long as the inherent safety provisions are maintained. (But may be realised if uncontrolled changes are introduced that undermine the inherent safety provisions.) If the suspended load is dropped the hazard event will be realised. The load is suspended by a number of cords, each of which represents a different defence or barrier. These cables are of different lengths; the shortest will carry the load, but if it should fail the load will transfer to the next shortest. The cords may also have different 'strengths' corresponding with their probability of failure. A typical arrangement would be a pressure control system backed up with a high pressure trip function, backed up in turn by a relief system. In normal operation the load is carried by the control system and the other cords are slack. It is only if the control system cord should fail that the load is placed upon the high pressure trip cord. If that cord should fail, or be disconnected by an override, the load will be placed upon the relief system cord. If all cords are compromised the load will be dropped and the hazard event will be realised.

A 'safety net' may represent conditional modifiers such as occupancy or probability of ignition etc.; this would only be invoked if the load was dropped, and models the possibility that the consequences might not be realised even if the hazard event should occur.

If a hazard is only present intermittently, so that an 'enabling condition' is required, this may be represented in the same manner, but with the load sat on a projecting support beneath the arch so that all the cords are slack; it is only when the support is removed that the hazard potential arises.

If a cord is too long, it will not provide any defence even if it is not itself compromised; it would be too late in restraining the load to supress the hazard. It would offer only the illusory appearance of a defence. This would be the case if a trip point was too close to the hazard for a trip to be timely (response time longer than the process safety time), or if an alarm was too late for effective intervention.

If protection provisions share some common elements, this may be represented by parallel cords suspended from a common cord (or vice-versa).

If a system such as 2 out of 3 voting is employed, this could be represented as three parallel cords each with a 50% load capacity, although this may be asking rather too much of a simple illustration.

The model is illustrated over the page. No claim of complete fidelity is made and there is nothing so very profound here, but it is hoped the model will be useful in conveying an understanding of relevant concerns.

(To view an animated illustration of the model, visit www.sissuite.com)

FUNCTIONAL SAFETY IN PRACTICE

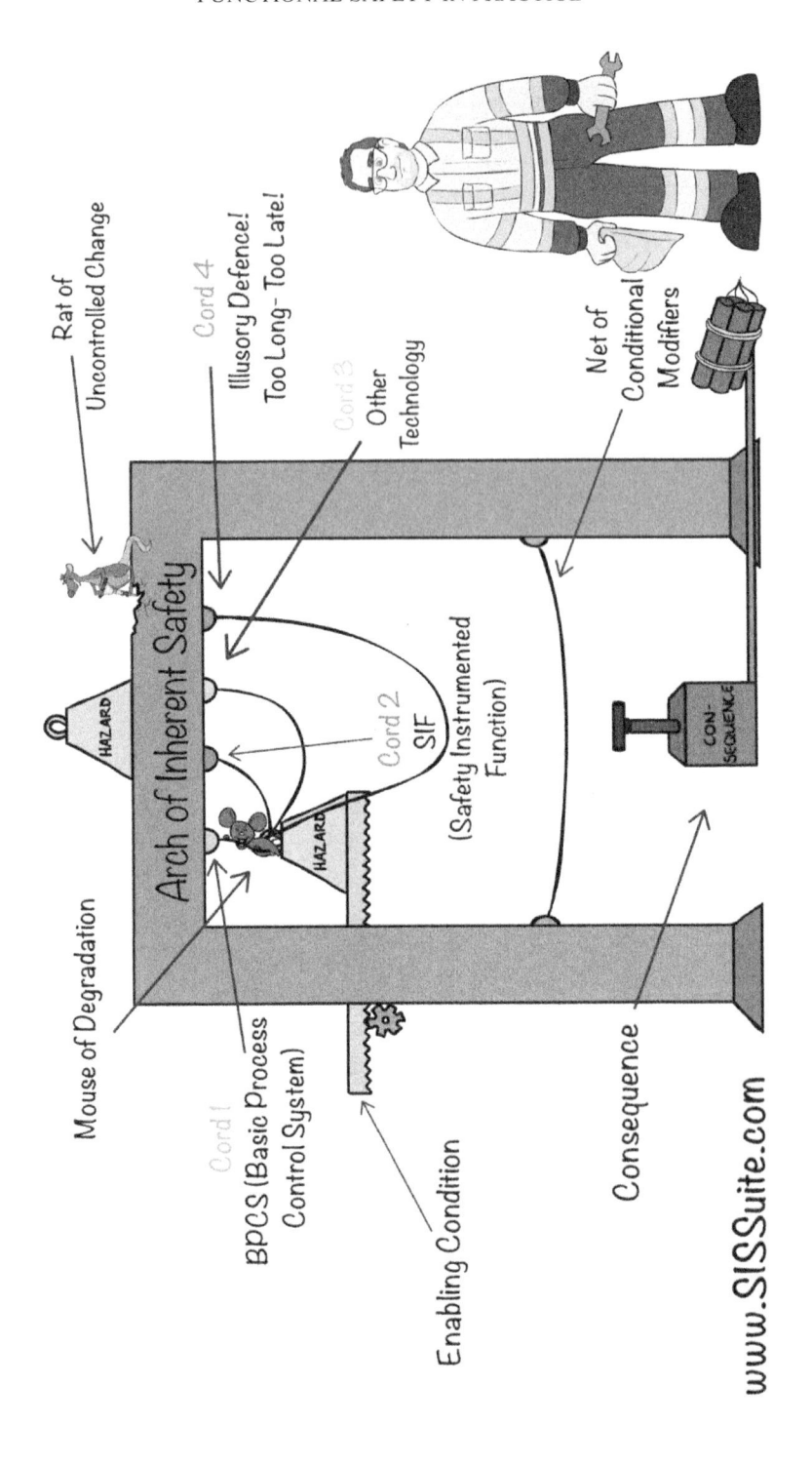

## 34. SECURITY: PHYSICAL & CYBER

If high integrity is required it is necessary to prevent unauthorized or inadvertent modification of the SIF. In that respect it is appropriate that physical access should be controlled and only granted to authorised personnel. The specific arrangements, both physical and procedural, should however remain practicable; if they are not, they are likely to be routinely ignored and will thereby undermine the wider safety culture discipline. Locked cabinets, locked rooms, logged access, and CCTV recording may all have their place. But simply saying 'everything SIF related must be locked' is unlikely to be a practicable approach. The 3 Ps; policy-procedures-practices, must be appropriately aligned. What needs to be locked? How? Who needs access? How will it be granted and controlled? How will it be revoked? How are security lapses or violations to be handled? All must be carefully thought through, properly communicated and managed. The first step is to identify all the security perimeters; where are the safety systems assets located, where are the access points?

What then of cybersecurity? The first point is that if you can decently avoid programmable logic solvers, you should, since it immediately eliminates many (not all – see below) avenues of cyber-attack which would otherwise require particular discipline in ongoing vigilance and management. What do I mean by 'decently'? I mean in recognition of the context in which the system will be deployed. Whatever the attractions of a programmable logic solver solution in terms of the project engineering, the ultimate concern should be its ongoing fitness-for-purpose in the context to which it will be deployed and key to this will be the capabilities of the duty holder during the operation and maintenance phase. Certainly there are many advantages in the flexibility of programmable logic solvers or their operational integration with the BPCS, but inevitably this very flexibility introduces vulnerabilities that must be managed. There is no once and for all design fix; continuing vigilance is required. If sustained management of these issues throughout what might be a 25+ year O&M phase are going to be difficult, then it may be better to avoid them in the first place. Of course, it may be that the particular requirements mean that a programmable solver is the only practicable (decent?) solution, in which case the important thing is to make sure the associated risks are properly assessed and addressed. Project and lifecycle planning should reflect this.

Avoidance of interconnectivity with other systems is an approach favoured by many; protection systems are 'air-gapped' from enterprise level systems and the BPCS. This does not guarantee cybersecurity however. Vulnerabilities may remain through USB ports and flash drives/mobile phones, the intermittent connection of maintenance computers, wireless connectivity, VPN and field devices. VPN may make data transfer secure, but if the 'message' (data packet stream) is itself corrupted then it offers no defence. There are apocryphal stories of engineering managers logging in from home through a VPN on a laptop infected by their kids' misadventures on the internet.

Note that an absence of a programmable logic solver does not eliminate all avenues of cyber-attack; there remain possibilities of attacks on sensor and final element subsystems. Many field devices are 'intelligent' and therefore susceptible to corruption or exploitation as a portal to higher system levels. Without a common programmable logic solver however it becomes difficult to engineer a co-ordinated attack on a facility. Without co-ordination it is difficult to make a dangerous attack that generates a hazardous event, as distinct from a disruptive attack that causes loss of production. A variety of security provisions may be claimed for individual devices, and concerns with wireless connectivity, for example, are well understood and provided for, but here is a key point; a particularly insidious threat is that of malicious re-ranging. This could cause a healthy and uncorrupted device to 'lie' about the state of the process, potentially causing process excursions beyond the operational design envelope. This could lead to demands on the protection systems and loss of plant availability, or if the protection systems are also compromised, a hazardous incident. Note that the device would remain healthy and uncorrupted – diagnostics will not help.

Field devices, whatever their security provisions, remain vulnerable to an insider attack. And if an insider is motivated to sabotage an operation there will be many ways to do it – and it need not be through a cyber-attack, it could literally be by 'throwing a spanner in the works'. The possibilities are legion.

Much is sometimes made of the potential for intelligence gathering through hacking into data systems, the notion being that threat agents could discover a plant's processing vulnerabilities, (essentially where to make a plant go 'bang'), and due diligence is off course required, but in considering defences against this potential the possibilities of an 'old school' approach through direct social interaction with employees might be overlooked. (Sex, drugs, ~~rock & roll~~, gambling and drink being the usual levers.) There is little point locking the front door if you are going to leave the back door open. An external attack via an insider proxy is particularly difficult to defend against; the insider might be persuaded to grant remote access or undertake an attack on behalf of an external threat agent.

Defences lie as much in the human resources domain as in the engineering one. We may engineer things so as to make unauthorised access difficult but appropriate administrative responses to behavioural cues are also required. Rules should be established about when individual access authority should be revoked e.g. departure, violation of security protocols, where disaffection is evident, drug/alcohol policy violations. This underlines the need for practicable security arrangements that do not promote routine breaches in order for employees to do what they need to.

## 35. SIL & CYBER SECURITY LEVELS (SL)

This chapter explores a possible approach to relating security level (SL) targets to the Safety Integrity Level (SIL) targets for protection functions implemented in Industrial Automation Control System (IACS) networks that could be subject to a cyber-attack. The IEC 62443 [1] standard introduces a range of Security Levels (SL) (1-4) which correspond with differing strengths of cyber-attack, the higher the level, the greater the strength of the attack (motivation, resources, sophistication) and the lower the likelihood.

It is tempting to draw parallels with the four safety integrity levels (SIL) familiar from the functional safety standard IEC 61508 and its derivatives, but there are dangers in this – there is no direct correspondence and the basis on which the levels are assigned, engineered and evaluated are very different.

The standard identifies specification of a target security level (SL-T) and evaluation of the achieved security level (SL-A) based on equipment capability (SL-C) and associated counter measures. Reference is made below to the SL-T and SL-A levels. Assessment is typically made of individual 'zones' within an IACS network consisting of so called zones & conduits.

It is acknowledged that quantification of cyber security risk is difficult, but there remains a notional separation of the Security Levels (SL) by an order of magnitude.

Designation of an SL-T of n implies that a threat at SL n+1 would be successful; there would be no risk reduction for that threat level. In this respect there is a key difference with Safety Integrity Levels (SIL) where a provision offering SIL n would still offer a risk reduction (albeit deficient) against a scenario that required a SIL n+1.

We might identify the nominal likelihood of attack at each SL as follows:

| Security Attack Level (SL) | Attack likelihood 1 in x target-years |
| --- | --- |
| 1 | >=1-<10 |
| 2 | >=10-<100 |
| 3 | >=100-<1,000 |
| 4 | >=1,000-<10,000 |

**Table 35.1** Nominal Likelihood Bands

If a zone was engineered with SL-A=n, the implication is that a successful attack on that zone should be tolerable as much as every $10^{n-1}$ years.

A likelihood of 1 in 100 target-years means that if there were 100 such targets, we would expect one would suffer an attack in any given year. This relationship between attack level and likelihood might be adjusted on the basis of the attractiveness of the target, with corresponding adjustments to the table above.

(An enterprise with a high public profile and the potential for high public service disruption might be deemed an 'attractive' target with a correspondingly higher likelihood.)

Note however, that many attacks will be revealed by anomalous system behaviour, allowing corrective action before a 'demand-in-anger' on a low demand Safety Integrity Function (SIF).

If we consider a low demand mode SIF, we may identify the tolerable successful <u>unrevealed</u> attack likelihood, regardless of the architecture concerned, as:

$$Tolerable\ Successful\ UNREVEALED\ Attack\ Likelihood = \frac{2 \times SIF\ Target\ PFD}{Test\ Interval}$$

This is simply back calculating (using the simple 1oo1 formula for Probability of Failure on Demand) from the target PFD to the corresponding unrevealed dangerous cyber-attack rate which would constitute a failure of the SIF.

At the SIL 2/3 boundary, if the test interval was 2 years, the tolerable successful unrevealed zone attack likelihood would be 0.001 (1 in 1000.)

If we assume 10% of attacks would be unrevealed this would correspond with a tolerable successful attack (revealed + unrevealed) likelihood at the SIL 2/3 boundary of 1 in 100.

$$Tolerable\ Successful\ TOTAL\ Attack\ Likelihood = \frac{2 \times SIF\ Target\ PFD}{Test\ Interval \times Unrevealed\ Fraction}$$

From the table above this would correspond with an SL of 2 at the SIL 2/3 boundary. (If the PFD target was less than 0.001, the test interval was less than 2 years, or the expectation was that more than 90% of attacks would be revealed, the SL target would be 1.)

If a SIF was operating in high demand/continuous mode, then there would not be the same prospect of corrective action before a demand-in-anger; the fact that the attack may be revealed is no longer relevant and testing would not help:

$$Tolerable\ Successful\ TOTAL\ Attack\ Likelihood = SIF\ Target\ PFH$$

At the SIL2/3 boundary this would be $10^{-7}$. Multiplying by $10^4$ for nominal PFH-PFD SIL band equivalence, this corresponds with a tolerable likelihood of 0.001 per target-year which would place the requirement at the SL 3/4 boundary; the SL target for the zone would increase by +1 over that for a low demand SIF of the same SIL with 10% unrevealed attacks.

If BPCS *control* and the associated SIF are within the same zone (pressure control and an associated high-pressure trip say), then they may both be compromised from a common cause attack, removing the potential for corrective action following an attack since the control failure may lead to excursion that would place a demand on the failed protection. From a cyber-attack perspective, such a SIF would correspond with the high demand mode situation; a successful attack would potentially place a demand on the disabled protection before a test could reveal the failure.

## SIL AND CYBER-SECURITY LEVELS (SL)

The tolerable failure rate is implicit in the PFD target and the demand rate used in determining this PFD target. (Note that this basis provides a driver for the separation of SIF and related BPCS control functionality into separate zones.)

$$Tolerable\ Successful\ TOTAL\ Attack\ Likelihood \\ = SIF\ Target\ PFD \times SIF\ Demand\ Frequency$$

If any *non-control* IPL is included in the zone, (typically an alarm) then this would be potentially susceptible to common cause cyber-attack together with any associated SIF. Assuming an aggregate magnitude risk reduction from any in-zone IPL of 'm', the overall target PFD for the in-zone protection (SIF + IPL) may be identified as:

$$Overall\ Zone\ Target\ PFD = \frac{SIF\ Target\ PFD}{10^m}$$

This may then be substituted as the target PFD in the formula above for tolerable attack likelihood:

$$Tolerable\ Successful\ TOTAL\ Attack\ Likelihood \\ = Overall\ Zone\ Target\ PFD \times SIF\ Demand\ Frequency$$

From these considerations, and in recognition of the limited scope for quantification in this field, for most practical purposes we may identify the following requirements for zone SL-T based on the attack likelihood table above:

| Zone Functionality | SL-T |
|---|---|
| Low Demand SIF | =SIL |
| High Demand/Continuous Mode SIF | =SIL+1 |
| LD SIF + Associated BPCS Control | =SIL+1 |
| LD SIF + Associated BPCS IPL | =SIL+1 |

**Table 35.2** SL-T assignment with Zone Functionality

If more rigour is preferred, the tolerable attack likelihood can be explicitly calculated from the target PFD, the test interval, and the estimate of the revealed fraction of attacks, this can then be used to look up the required SL from a nominated SL v likelihood interval table calibrated appropriately for the operation concerned.

The SL-T for a zone would correspond with the highest value for the functionality implemented within that zone. It is recognised that IEC 61508 allows more than one IPL claimed for a given SIF implemented in BPCS, but the requirements for independence would militate against them being in the same zone and this would limit the claim for BPCS IPL to one per zone.

If the PFD/PFH target is already just met by the random hardware failure rate specifications, then the implication of using target PFD/PFH to identify the tolerable attack likelihood is that in combination the two sources of failure (random hardware failures AND cyber-attacks) would lead to the target hazard rate being exceeded. In practice, given the uncertainties associated with the estimates involved, the typical conservatism used and the SIL/SL band widths, this may well not be a concern; the approach may be deemed proportionate. If a more rigorous approach is required, the prevailing gap between SIF PFD/PFH and the target might be used so that in combination the risk from random hardware failure AND cyber-attack still meets this target. Enhanced reliability, or, for a low demand function, more frequent testing, would expand this gap leaving 'more room' for cyber-attacks thereby possibly reducing the SL target. An iterative approach to target specification might be anticipated.

For *business* impact the risk from disruption from cyber-attack may be compared with that due to plant breakdown. Recovery from a cyber-attack may take several days and may oblige a halt in production. Plant downtime can quickly lead to high lost opportunity costs but there is little point in strengthening defences to reduce the risk of cyber disruption far below the risk associated with plant breakdown. If the likelihood of a 3-day shutdown due to plant breakdown is 1 in 10 years say, there is little point in targeting 1 in 100 years for a 3-day shutdown due to cyber-attack unless the additional engineering burdens/costs are trivial.

A notional correspondence between downtime likelihood and cyber-attack likelihood may be used to identify the SL-T:

# SIL AND CYBER-SECURITY LEVELS (SL)

| Unscheduled Downtime Likelihood (1 in x years) | Corresponding Cyber Attack Likelihood Interval Band (SL-T) |
|---|---|
| 0.5 | 0 |
| 1 | 1 |
| 2 | 1 |
| 5 | 1 |
| 10 | 2 |
| 50 | 2 |

**Table 35.3** SL-T assignment with Downtime Likelihood

From consideration of the likelihood of a given breakdown downtime period, a site-specific business risk calibration may be identified. For example:

| Downtime Duration (days) | Unscheduled Downtime Likelihood (1 in X years) | Corresponding Cyber Attack Likelihood Interval Band (SL-T) |
|---|---|---|
| ½ | 0.5 | 0 |
| 1 | 1 | 1 |
| 3 | 2 | 1 |
| 7 | 5 | 1 |
| 14 | 10 | 2 |
| 30 | 50 | 2 |

**Table 35.4** Site Specific Business Risk Calibration

From consideration of these requirements the highest SL-T can be identified for each zone.

**References:**

1. IEC 62443: Security for industrial automation and control systems.

## 36. COMMON CAUSE & BETA FACTORS

We understand that the calculations for PFD are based on random hardware failure. In which case it might seem odd that we are required to include a common cause factor when evaluating redundant systems, since the fact of there being a common cause suggests a systematic rather than random influence. Full details of the considerations that explain the apparent anomaly may be found in Annex D of part 6 of the standard. A telling point is the scope for second order effects; the higher the unreliability and complexity of a system, the higher the requirement for maintenance and testing, and the higher the potential for the introduction of systematic susceptibilities into the design and installation.

Common Cause Failures (CCF) occur when multiple failures arise from a single shared cause. The multiple failures may occur simultaneously or over a period of time. For example, the environmental conditions may cause two channels to fail for the same reason but with the failures a week apart. The separation in time may allow diagnostic detection and intervention before the second failure arises.

Common Mode Failure is a particular case of CCF in which multiple equipment items fail in the same way (mode).

So, when using redundant channels, the questions arises of the beta factor to be used when evaluating the SIF PFD using the beta factor model to account for the common cause failure possibilities.

IEC 61508 identifies two values for beta factor:

> $\beta$ is the common cause failure factor for *undetectable* dangerous faults
>
> $\beta_d$ is the common cause failure factor for *detectable* dangerous faults (i.e. where diagnostics are employed)

Both are used in the formulae for PFD from the reliability block diagram approach as given in Annex B of Part 6.

A full analysis in accordance with the informative Annex D of Part 6 of IEC 61508 is an option (but can be quite challenging in a number of areas), but as a simple conservative approach the following values (for both $\beta$ and $\beta_d$) may be adopted for SIFs that are appropriately engineered to be SIL compliant. These values are identified as conservative since they are based on the *minimum* scores of table D.4 (and the architectures from table D.5). (The beta factor is not of course applicable for 1oo1 or 2oo2 architectures since there is no redundancy).

| MooN | $\beta$ and $\beta_d$ Logic Subsystem | Sensor/Final Element Subsystem |
|---|---|---|
| 1oo1 | - | - |
| 1oo2 | 5% | 10% |
| 2oo2 | - | - |
| 2oo3 | 7.5% | 15% |
| 1oo3 | 2.5% | 5% |
| 2oo4 | 3% | 6% |

**Table 36.1** Conservative Beta Factor Values

The approach in Annex D uses a table of nominal score values against different aspects (30+) that are weighted and summed and then banded to provide a value for the beta factors. The scores in the table are 'based on engineering judgement'. Once again, given the broad uncertainties that run throughout functional safety concerns, I am sceptical whether the elaboration and attempted degree of discrimination is warranted. As another approach we might adopt the ranking method as employed for proof test coverage in Chapter 37.

We may identify those factors that are likely to correlate with beta factor:

**Subsystem Type:** Sensor, logic or final element. Field equipment being typically more exposed than equipment housed in panels in auxiliary rooms.

**Systematic Integrity:** The better the defences against systematic failures, the better the defences against common cause failure. Identified through device systematic capability and compliance with Part 2 Annex A *'Techniques and measures for E/E/PE safety-related systems – control of failures during operation'*.

**Diversity:** A given systematic integrity may be achieved with or without diversity; we may anticipate improved defence against common mode failure if diversity is employed.

**Diagnostic Effectiveness:** The better the diagnostic coverage and the better the frequency with which the diagnostics operate, the better the defence against a common cause failure.

Adapting the intervals from tables D.2 and D.3 from IEC 61508-6 to our purpose and categorising diagnostic effectiveness as High/Medium/Low:

| Diagnostic Coverage | Diagnostic Test Interval |  |  |  |
|---|---|---|---|---|
|  | <2 hours | 2 hours - 2 days | 2 days - 1 week | >1 week |
| ≥99% | H | H | M | L |
| ≥90% | H | M | L | L |
| ≥60% | M | L | L | L |

Table 36.2 Diagnostic Effectiveness Categories for **Sensors/FE**

| Diagnostic Coverage | Diagnostic Test Interval |  |  |
|---|---|---|---|
|  | <1 min | 1 min - 5min | >5 min |
| ≥99% | H | H | M |
| ≥90% | H | M | L |
| ≥60% | M | L | L |

Table 36.3 Diagnostic Effectiveness Categories for **Logic Solvers**

These factors then form the four dimensions of 'common cause susceptibility space'.

IEC 61508 identifies a maximum base (1oo2) beta factor of 10% for sensors and final elements and a minimum of 1%, with corresponding values for logic of 5% and 0.5%.

Although the standard here treats sensors and final elements as a single category, I question this. Valves might be sourced from different manufacturers, but they all rely on the actuation of a closure member in the process flow stream. Contactors/breakers all rely on similar principles for their operation. These elements may well be predominantly susceptible to the same degradation mechanisms. I suggest final elements do not have the same degree of diversity as sensors. A given sensor duty might be fulfilled by a wide variety of technologies. Consider flow for example: Coriolis, vortex,

FUNCTIONAL SAFETY IN PRACTICE

electromagnetic, orifice, turbine, thermal, pd, etc. The same cannot be said of final elements. For this reason, I elect to increase the maximum for final elements to a nominal 15%. We now have the upper and lower bounds to our space.

We may also postulate that diverse redundancy at one level of systematic integrity has nominally the same susceptibility as identical redundancy at the next higher level of systematic integrity.

We may then rank positions within this space:

| Subsystem Systematic Integrity | 3: (SC2 Diverse/SC3 Identical) | | | 2: (SC1 Diverse/SC2 Identical) | | | 1: (SC1 Identical) | | |
|---|---|---|---|---|---|---|---|---|---|
| Diagnostic Effectiveness | H | M | L | H | M | L | H | M | L |
| Logic | 0.5% | 0.5% | 1.25% | 0.5% | 1.25% | 2.5% | 1.25% | 2.5% | 5% |
| Sensor | 1% | 1% | 2.5% | 1% | 2.5% | 5% | 2.5% | 5% | 10% |
| Final Element | 1.25% | 2.5% | 5% | 2.5% | 5% | 10% | 5% | 10% | 15% |

**Table 36.4** Base (1oo2) Beta Factors

This table then offers a base (1oo2) value for both β and βd. **(Use the low effectiveness columns for β and the assessed effectiveness columns for βd)**. The appropriate base values may then be modified for the architecture employed using the values from table D.5 of IEC 61508:

| MooN | Factor |
|---|---|
| 1oo3 | x0.5 |
| 2oo3 | x1.5 |
| 2oo4 | x0.6 |

**Table 36.5** Multiplication Factors for Architectures other than 1oo2

So, for 2oo3 *sensor* redundancy without diversity (*identical*), with *medium* effectiveness diagnostics, and *systematic integrity* that supports SIL 2, the β would be 5% x 1.5 = 7.5% and the βd, 2.5% x 1.5 = 3.75%.

Note that although a result like '3.75%' may appear to offer precision, (having three significant figures), this is merely an illusion. The figures are nominal only, having been generated by an essentially qualitative evaluation across the four dimensions with a notional proportionality with each.

Note that anything other than low effectiveness **may only be claimed if**:

a) the SIF (not just a channel) is triggered on detection of a fault (as would be the case with 1oo2/1oo3 sensors), or

b) the process is shutdown or a repair effected within the claimed diagnostic interval.

The common cause factor is applied to redundant parallel branches in the reliability block diagram. These branches may well include multiple elements in series e.g. (transmitter + barrier + trip amplifier), or (relay + SOV + ESD valve).[a] For most practical purposes the beta factor nomination may be made on the basis of the dominant element; typically the transmitter or the valve, and where diversity is employed may be conservatively nominated on the basis of the least capable of the dominant elements from the parallel branches.

---

[a] In SISSuite the diagnostic coverage, SFF, type and SC is automatically calculated for such a chain in a 'supertag'.

## 37. PROOF TEST COVERAGE NOMINATION

The rigorous approach to determination of proof test coverage is to identify the dangerous failure rate associated with each dangerous failure mode and to consider which modes would be revealed by the test method and then calculate the corresponding percentage of the total dangerous failure rate.

In truth however, such rigour is a fantasy. There are such broad uncertainties in the failure rates and the identification of the failure modes in any given context, that attempts to determine precise percentages are usually wrongheaded. Particularly when considered alongside the broad uncertainties in other aspects of functional safety.

Discussions concerning proof test coverage are typically framed in the context of PFD calculations, since it will bear on the frequency with which testing is required. But PFD calculations are predicated on *random* hardware failure rates and testing also has a role in detecting *systematic* failures due to design and installation issues or maintenance interventions.

So, how may we proceed? The important thing is to recognise that most test methods will provide less than full coverage. 100% coverage is typically very difficult to achieve. Many assume that a full stroke test of a valve with the plant shutdown will provide 100% coverage of the final element, but will the valve still operate when the plant is operational (designated 'live' in the following table), and the pipework stressing has changed and there is a significant pressure drop across the closure member? Your assumed 100% may in fact be only 99%, (or 90%, or 70%, or...). One possibility in order to enhance coverage is to use a 'real' demand by driving the process to the trip point, but there are difficulties with this; what if the protection did not operate - you may have introduced the very hazard you were trying to prevent! This approach may be used, but it needs careful consideration of how it is to be managed appropriately.

The implication of less than full coverage is that at some point some extraordinary intervention will be needed, typically an overhaul/replacement or a special test/inspection provision. The aim being to establish confidence that the element has been effectively returned to the 'as new' condition and that there are no residual unrevealed dangerous failures. Now the exact period with which this extraordinary intervention is to be undertaken is less of a concern; as long as the less than perfect coverage is recognised and the residual growth in PFD is not left indefinitely. The simplified PFD equations assume 100% coverage but the more advanced calculations include a proof test coverage factor and will identify the average PFD across a full cycle of partial tests and the point at which the asset is returned to the 'as new' condition.

Rather than attempt a rigorous calculation of coverage, which can typically only give an illusion of accuracy, we may categorise test approaches and rank them in terms of the coverage achieved; essentially in terms of 'good', 'better', 'best'. It is clear that some test approaches will achieve less coverage than others because they will form a subset of broader tests. We also know that the coverage will lie between 0 and 100%! From these considerations, we may assign nominal coverage percentages to typical approaches to correspond with their relative ranking. There is little enough science here, but given the broad uncertainties, that 'little' is sufficient for our purposes; the rankings are real, the constraints are real; worries about precise percentages would be entirely misplaced.

The values in the table are presented as a starting point – if you identify any particular considerations that may impact on coverage then these figures can be revised; they are not offered as 'gospel' but as nominal figures that might be used for the want of anything better. No claim of rigour is made, but they will help drive appropriate behaviours and proper consideration of test provisions. If you have credible specific data for the equipment as deployed in your installation, then you should use that, but that would be unusual. The table might be expanded, and further figures interpolated to cater for particular equipment types or installation circumstances. We distinguish between a 'functional test' that checks the protection operates and a 'proof test' that checks that there are no unrevealed dangerous failures in any element. If there is a 1oo2 system for example, it is not enough to know the protection operated, you need to know that both channels are healthy, or a dangerous unrevealed failure could persist in one channel. Note that observing fall off in downstream flow/pressure would not prove BOTH valves in a 1oo2 system have operated; successful operation of one might mask the failure of the second.

The table distinguishes between fail-safe open (FSO) and fail-safe close (FSC) valves and between those that need tight shut-off to effectively suppress the hazard. Note that it is unusual for Tight Shut-Off (TSO) to be required to supress a hazard; more typically it would be sufficient to effectively throttle the flow 'to a trickle'. TSO is normally a requirement for operational reasons rather than hazard suppression. Avoid calling for TSO as part of your Safety Requirement Specification (as distinct from the procurement specification) if it is not a requirement for safety.

If a valve needs to operate quickly because there is a relatively short process safety time, (where the PST is <10 x stroke time say), then there is additional scope for degradation of the protection function through deteriorating stroke times which may then make up a greater proportion of the total failure rate.

For out-of-line measurement devices that make use of process connection tappings which might be blocked or otherwise compromised in some way, consideration must be given to proving of these connections as well as the measurement device itself.

Measurement comparison might validate a measurement at a given point, but this would not guarantee that a dangerous discrepancy would not arise at the trip point. Coverage will naturally be better if the comparison is between diverse devices rather than two measurements using the same technology.

For logic solvers that handle multiple functions, at plant shutdown you will typically see the outputs in their safe (tripped) state. The inputs may be exercised, and the logic solver input status response checked. On this basis, the safe state of the logic solver will have been comprehensively checked. It would be a relatively obscure fault that would spuriously drive an output into an unsafe condition – provided of course that there has been no interference with the configuration. A full positive path test might reveal these more obscure fault possibilities but can be difficult to perform and might involve significant levels of disturbance to the installation.

Contactors and breakers are much more straightforward to check by confirmation that they are in the safe state with the associated equipment de-energised, but it is important to check that the safe state is not due to some other interruption in supply, and that the associated contact in the stop circuit is operating correctly.

Inspection may reveal degradation of the equipment that is yet to manifest as an actual failure and so a nominal 5% is assigned to close inspection. 'Inspection' in this context refers to a formal close external examination to check for degradation due to e.g. impact, corrosion, vibration, contamination, unauthorised modification etc.

What of service conditions? Well if there is a particular failure mechanism that will be exacerbated by the service, then a corresponding adjustment to the coverage factor might be appropriate, but this might be more simply addressed by using a reduced overall failure rate claim for elements on such services, and a reduced useful life, with a corresponding requirement to overhaul/replace earlier than would otherwise have been the case. The equipment should be selected to be fit-for-purpose in the knowledge of the intended service and should have a specification that will support that service. This is the underlying assumption in the notion of returning a device to the 'as new' condition.

Notes to the Table:

a) This check may only be claimed with a single valve; failure of redundant valves might be masked.
b) 'Shutdown' meaning the process is not at operating pressure and temperature; as distinct from 'live' conditions.
c) A 'safe state check' is simply a check that the valve (or drive) has moved to the correct position (state) without observation of the operation. Note that observation of the stroke may not reveal a sheared shaft/linkage; it is only through observation of the flow that we can be sure that a valve is throttling/venting/draining correctly (and then only if we can be sure that it is not some other element that is influencing the flow e.g. pump stop).
d) Process simulation would typically be by use of a process calibrator e.g. a pressure calibrator.
e) Amplifier injection would typically be a low-level signal injection to the amplifier e.g. for electromagnetic/vortex flowmeter or the use of a test button on a tuning fork switch amplifier.
f) Output drive would typically be through forcing the output of an intelligent device through a device configuration interface.
g) If the sensor is remote from the process proper e.g. with an impulse line connection to a pressure/level transmitter, then this process connection itself might be a source of failure and should also be tested and inspected.

PROOF TEST COVERAGE NOMINATION

| FE Subsystem Channel | | FSC | FSC TSO | FSO |
|---|---|---|---|---|
| Live TSO check (Timed/Untimed) (a) | OR | - | 95/80 | - |
| Live Throttle/Vent check (Timed/Untimed) (a) | | 95/80 | 80/65 | 95/85 |
| Live stroke check (Timed/Untimed) | | 90/75 | 70/55 | 90/80 |
| Shutdown (b) stroke check (Timed/Untimed) | | 85/70 | 60/45 | 85/75 |
| Safe State Check (c) | | 65 | 40 | 70 |
| Partial Stroke | | 60 | 35 | 60 |
| Inspection | AND | 5 | 5 | 5 |
| Untimed with Short PST | AND | -15 | -15 | -15 |

| Contactor/Breaker | | | | |
|---|---|---|---|---|
| Safe State Check | AND | 75 | | |
| Stop Circuit Safety Contact | AND | 25 | | |

| Logic Solver | | | | |
|---|---|---|---|---|
| Safe state check | OR | 85 | | |
| Full positive path test | | 95 | | |
| Inspection | AND | 5 | | |

| Sensor Subsystem Channel | | Tx | | Switch | |
|---|---|---|---|---|---|
| | | In-Line | Remote | In-Line | Remote |
| Forced process excursion | OR | 95 | 95 | 95 | 95 |
| Workshop/Lab + In-situ Output Drive Check | | 95 | 75 | 95 | 75 |
| Process simulation (d) | | 80 | 60 | 80 | 60 |
| Measurement Comparison (Redundant) | | 50 | 30 | - | - |
| Measurement Comparison (Diverse) | | 70 | 50 | - | - |
| Amplifier Injection (e) | | 60 | 40 | 75 | 50 |
| Output Drive (f) | | 40 | 20 | 10 | 10 |
| Process Connection Test if NOT Excursion Test (g) | AND | - | 20 | - | 20 |
| Inspection | AND | 5 | 5 | 5 | 5 |

**Table 37.1** Nominal Proof Test Coverage Percentages.

## 38. MULTIPLE SIF LAYERS

Imagine you are doing a LOPA study to identify the SIL target for a new SIF, As part of the analysis you identify an existing separate and **independent** SIF (different sensors/logic/FE) as an independent layer of protection (IPL) - let's call it SIF 'A'. (Or possibly you propose one as a design solution.) Assume it offers SIL 1 with a risk reduction of 20. Your LOPA identifies a risk reduction requirement from the new SIF ('B' say) of another 20 (to keep the sums simple). So, we have:

SIF A: SIL 1 RRF 20

SIF B: SIL 1 RRF 20

The total SIF RRF requirement is 20 x 20 = 400. This is SIL 2 territory.

If you implemented this as one 'big' SIF (SIF C) the target would be SIL2 with corresponding requirements for systematic integrity and hardware fault tolerance.

If you make SIF C 1oo2 by cross connecting the subsystems from what would have been SIF A and SIF B you will potentially enhance the integrity over what would have been achieved with two 1oo1 SIF.

Two separate SIFs like this;

## MULTIPLE SIF LAYERS

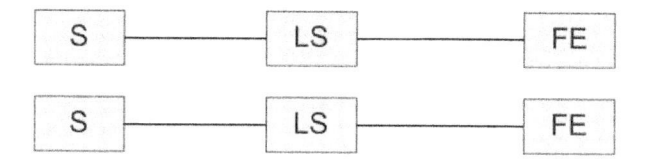

```
[S] ──── [LS] ──── [FE]
[S] ──── [LS] ──── [FE]
```

are, **all else being equal**, not as good as a scheme like this:

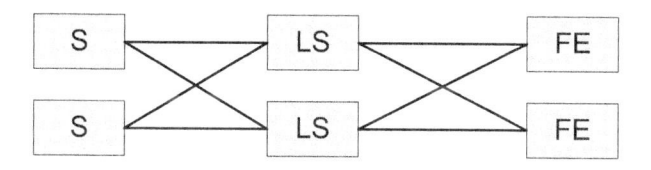

```
[S] ╳ [LS] ╳ [FE]
[S] ╳ [LS] ╳ [FE]
```

Since, for example, a sensor failure in SIF A and an FE failure in SIF B (or vice versa) would disable all the protection. But the equivalent failures in SIF C would still see the protection available. Note also the two separate SIF may still have potential common cause failures (not least through operational and maintenance activities). A simple beta factor model can be applied to the two SIF to calculate a combined RRF. Using a 10% beta factor, this yields RRF 120 (not the 400 you might have hoped for).

[With an annual test the equivalent (1oo1) dangerous failure rate for each SIF would be 0.1/year. If 1oo2 with 10% beta factor, the simple PFD calculation then yields (0.1 x 0.1)/3 + 0.01/2 = 0.0083 or RRF 120]

The standards do allow the use of multiple SIF to satisfy an overall safety integrity requirement, but if independence is to be claimed they require demonstration that '...*the likelihood of simultaneous failures between two or more of these different systems or measures is sufficiently low in relation to the required safety integrity*'. (61508-1 7.6.2.7) Requirements then also include diversity of both functional approach **and** technology, and that the SIF do not share parts, services, support or O&M procedures. (So quite a tall order in practice). IEC 61511 is less explicit, but the requirement stands (naturally). [see 61511-1 9.2.6 and 61511-2 A9.2.6, which explains the provisions relate to '...*higher levels of risk reduction (e.g. greater than $10^3$)*'].

If independence is **not** claimed(demonstrated) then full consideration of dependencies and common cause potential is required.

So, there are three possibilities:

i) one big SIF
ii) two independent SIF
iii) two non-independent SIF

If **non-independent** you should specify a systematic integrity in support of SIL2 (for my example) for both the SIF and include an appropriate common cause calculation in the assessment of the overall PFD/RRF for the combined SIF. If SIF A (or whatever) is already existing, it might be difficult to retrospectively upgrade it to formally meet the IEC 61508 SC 2 requirement, but if there is good experience it might well be appropriate to make a claim on the basis of IEC 61511 and 'prior-use', otherwise an ALARP case might be made for keeping the SIF as it is. There would however remain an obligation to test the SIFs appropriately to meet the overall RRF target.

If **independent**, a systematic integrity in support of SIL 1 would be sufficient. (But remembering the diversity and other requirements of independence.)

If using the **'big SIF'** scheme, and the two channels within any subsystem were sufficiently diverse, you might argue that in combination, they meet the overall requirement of SC 2 even though they were separately assessed as SC 1 (per 61508-2 7.4.3.3).

If the two SIF were completely diverse and segregated etc.,[a] then there may be a case for preferring this over 'one big SIF', since it might be argued that their integration in that way would make them more susceptible to common cause failure (particularly through operation or

---

[a] The use of a simple Beta model is questionable if the SIFs are 'completely diverse and segregated etc.' and therefore do not share common failure modes, but it serves to provide the ranging shots for my illustration.

maintenance interventions). If, as a 'ranging shot', in recognition of the diversity, we use a beta factor of 0.5% (rather than the 10% used earlier), we would have (0.1 x 0.1)/3 + 0.0005/2 = 0.00358 or RRF 279.

Interestingly, if we assume a nominal distribution of the overall 0.1/year failure rate as due to 35% sensor, 15% logic and 50% final element and, per the second scheme, we use a 1oo2 configuration for each subsystem with a 5% beta factor, we end up with essentially the same result: RRF 271.

So, we would need to establish a degree of independence that offers better than the equivalent of 0.5% beta factor to improve on the 'big SIF with 5% β' proposition.

This example illustrates why we should not be over reliant on the numbers game. The calculations are a potentially helpful tool to help us probe the options, but they use models that do not necessarily capture all the nuances of a situation and they should not take precedence over pragmatic considerations about the engineering and the context in which the protection will operate. Unless you can demonstrate true independence, the requirement for systematic integrity that supports SIL2 would remain. Any of the schemes would comfortably meet the RRF 400 requirement with 6 monthly testing.

# 39. HUMAN FACTORS

There is a whole literature devoted to human factors; my very limited aim here is to consider the key considerations from the perspective of functional safety only. The potential for human error may be said to run throughout the safety lifecycle, but the processes of verification and assessment provide defences to much of this. The particular vulnerabilities lie:

a) *directly* in a failure to properly execute operation and maintenance procedures such as:

- Proof testing
- Maintenance/repair
- Management of change

that may disable or compromise the protection by leaving SIF elements isolated or bypassed or introducing defects (e.g. flawed calibration or configuration).

b) *indirectly* through flawed assumptions and estimates in relation to human error in terms of initiating event frequency or independent protection layers, particularly alarm response.

Claims for effective human intervention to prevent or mitigate a hazard are typically quite limited; a PFD of 0.1 (RRF 10) will usually be allowed provided the circumstances do not *militate* against this. So, there needs to be:

- the means to recognise what action is needed
- the means to perform the required action
- sufficient time to recognise and act upon the need

If the ergonomic arrangements will not support this then the claim should be disallowed. Note also that a competent operator who is not distracted or overloaded with other tasks would need to be available.

What would constitute 'sufficient time'? Much will depend upon the particulars but I suggest anything less than 10 minutes would be immediately questionable. (The UK civil nuclear sector does not allow a claim where action is required inside 30 minutes). Even 10 minutes would require a demonstrably strong 'signal-to-noise' ratio to be confident the alarm would not be missed, and a straightforward and simple means of intervening to be sure the appropriate action would be executed competently. If the signal is very strong and clear and the action is so very straightforward, it invites the question; why not automate it?

Alarms must be appropriately prioritised and annunciated in a manner that means they will not be lost in an 'alarm flood'. Classically the approach has been to use 'hardwired' annunciation. A defined (documented) response procedure and corresponding training are also required.

Claims of better than a PFD of 0.1 would typically require particular substantiation, and may require formal task analysis and evaluation of human performance influencing factors.

For initiating events, an approach sometimes used is to identify the frequency of an operation and then postulate an error rate (PFD) in execution:

$$IE_f = OP_f \times PFD_{execution}$$

This needs some care however; the normal PFD for an alarm and associated response is sometimes propagated into the calculation of initiating event frequency, but failure to respond to an alarm is a sin of omission in circumstances where there may be other imperatives competing for attention. Errors in executing a routine operating procedure are typically much less likely. The operator's attention is already engaged with the task, it is practiced, there may be contrary indications if a mistake is made, there may be opportunity for corrective action. However, it must be recognised that for many tasks there will be multiple opportunities for error, possibly of different type, especially if the task is long and/or complex.

The table below (drawn from Kirwan, *A Guide to Practical Human Reliability Assessment,* 1994) is widely cited, and gives some indicative values showing the range that might be invoked. This table should not be regarded as definitive; it certainly does not cover all circumstances. Context is key when estimating human error probability.

| Description | Error Prob. |
|---|---|
| General rate for errors involving very high stress levels | 0.3 |
| Complicated non-routine task with stress | 0.3 |
| Supervisor does not recognise the operator's error | 0.1 |
| Non-routine operation, with other duties at the same time | 0.1 |
| Operator fails to act correctly in the first 30 minutes of a stressful emergency situation | 0.1 |
| Errors in simple arithmetic with self-checking | 0.03 |
| General error rate for oral communication | 0.03 |
| Failure to return the manually operated test valve to the correct configuration after maintenance | 0.01 |
| Operator fails to act correctly after the first few hours in a high-stress scenario | 0.01 |
| General error of omission | 0.01 |
| Error in a routine operation where care is required | 0.01 |
| Error of omission of an act embedded in a procedure | 0.003 |
| General error rate for an act performed incorrectly | 0.003 |
| Error in simple routine operation | 0.001 |
| Selection of the wrong switch (dissimilar in shape) | 0.001 |
| Selection of a key operated switch rather than a non-key operated switch | 0.0001 |
| Human performance limit: single operator | 0.0001 |
| Human performance limit: team of operators performing a well-designed task, very good performance shaping factors | 1E-6 |

**Table 39.1** Indicative Human Error Probabilities.

If an uncorrected mistake would lead to an alarm or incident then there is a ready measure of probability of failure per occasion. If a given operation was performed a 100 times a year and the postulated probability of failure was 0.1 then we would expect 10 incidents/alarms per year. This provides an immediate check on the plausibility of our estimates.

Our equation might be expanded:

$$IE_f = OP_f \times PFD_{execution} \times PF_{error\ recovery}$$

Many errors might be fail safe or fail 'no effect' rather than dangerous and our concern is only with the dangerous errors.

$$IE_f = OP_f \times dangerous\ PFD_{execution} \times PF_{error\ recovery}$$

This expanded equation might explain any apparent disconnect between unduly simplistic estimates of initiating events due to operator error and the actual frequency of alarms/incidents experienced.

Formal techniques such as HEART (Human Error Assessment and Reduction Technique) or THERP (Technique for Human Error-rate Prediction) provide probabilities for human error based on the nature of the task and the conditions under which it is performed. A full analysis will consider all the tasks within a given operation (procedure) to identify the aggregate risk of particular hazards arising.

As with other quantitative approaches in functional safety, the main value in formal assessment may lie in probing sensitivity to human factors, and identifying how procedures, training and assessments may be enhanced, rather than attempting to establish an absolute value for the likelihood of error.

## 40. OVERRIDES & RESETS

It is all very well engineering a high integrity SIF, but if the function can be readily overridden then it does not really offer high integrity; it might look the part, but it may not fulfil its promise.

A classic *faux pas* is to have a beautifully engineered override key switch panel in the control room, only to have the inspector come around and discover all the keys stored in the switches! This immediately demonstrates a lack of proper management control. There should be a defined procedure for management control of overrides.

The Part 6 Annex B calculations for PFD include 'Mean Repair Time' and 'Mean Time To Restore', which thereby account for the downtime for the SIF whilst repairs are undertaken (and overrides are in place) – the assumption being that the process hazard remains present. Many process operations would not continue to operate if a SIF was known to have failed, in which case a repair time of zero may be legitimately used in the calculation – regardless of how long it takes to repair the SIF.

Note that if a voting system such as 2oo3 is used, the application of an override to one initiator would degrade the function to 2oo2. It may be appropriate to reconsider testing provisions to return the PFD to the specified level. Leaving (or forcing) a faulty sensor in (to) a failsafe mode would mean a 2oo3 function operating as 1oo2: this would in fact mean that safety integrity was slightly enhanced, albeit with an increased risk of a spurious trip.

If an override of a function is to be employed, it usually makes more sense to override the initiator rather than the final element (so that the final element remains available). Particularly if the final element is used by more than one SIF or if it is part of a general shutdown associated with the operation of a manual ESD.

Overrides e.g. of a low flow trip, may be essential to allow start-up of a plant. The difficulty comes if these overrides should remain active post start-up. One approach is to alarm their continued application, prompting their removal by the operator. Another is to have automatic removal after a specified time has elapsed, leading, of course, to a plant trip if the running conditions have not been satisfied. A further option is automatic removal once an appropriate threshold has been passed – this may be the preferred approach if the override is only required for a short transition on start up or if the hazard only arises beyond that threshold. If possible the automatic override duration should be less than the process safety time at the start up condition.

Note that the SIL equipment requirements extend to override elements, since their failure or maloperation may disable the protection.

I have an aversion to the alarm technique; there may be many other concerns competing for the operator's attention and the removal of an override might not be seen as a priority. If the plant is in trouble, and the operator is correspondingly stressed, there may well be a tendency to disregard the alarm just at the time when the protection might be needed!

Another concern is the possibility that the operators might fall into the practice of using an override to avoid difficulties rather than having the underlying design/process issue addressed. If the process is not in a steady state and liable to excursions, there will be the temptation to apply overrides until the situation stabilises, but these may be the very circumstances under which the protection is required. If the process is not well controlled during a transition phase, the operator's attention may be elsewhere and it might be unreasonable to claim operator intervention as a defence. If a transient breach of a trip point is acceptable then it would be better to engineer a delay timer. The techniques employed for more intelligent alarming might be also deployed to plant trips to provide a greater degree of discrimination in their executive action. These considerations underline the need for proper management control of the application of overrides. If they are always available for use at the operator's discretion then there is little that may be claimed in the way of management control.

Override of groups of initiators or final elements (by a single switch/operation) needs particular care to make sure that just the appropriate functions are overridden and that protection that should remain available is not inadvertently disabled.

This discussion is in relation to engineered overrides, but the same considerations relate to *ad hoc* overrides ('frigs') such as the use of 'jumpers' (electrical or pneumatic) that might be used to disable protection provisions. The application of any such overrides should be controlled through formal Management of Change procedures. Of particular concern is the management provision for their review and removal. Jumpers should be suitably conspicuous (1m of fluorescent orange please, not 3cm of muddy brown) and managed through a register; count them all in, count them all out.

Security is a further concern, I have known installations in which operators, having observed the techniques employed by their instrumentation colleagues, have learnt to have a small jumper wire to hand on the night shift to overcome a 'local difficulty' and effect a reset; 99 times out of a hundred their ploy might be a successful manoeuvre, but being an uncontrolled modification they might come very seriously unstuck on the hundredth occasion (or the first!). High integrity protection functions should be housed in secure cabinets/rooms to prevent unauthorised (even though well-intentioned) interference.

Remember; a SIL 3 SIF with a SIL 0 override is in fact a SIL 0 SIF!

Resets also need some care, the systems should be engineered so that a reset command does not inhibit the protection; it should only make the final elements available again if the initiators are healthy. Some functions may not need a reset command operation; it may be acceptable that the protection returns to the healthy state ('auto resets') as soon as the initiator(s) go healthy. Whether the final elements should automatically return to the normal operating state upon a protection reset will depend upon the particular nature of the process and the final element concerned. For example, would you want a drive to start as soon as the protection is reset, or would you want it to only start when commanded by the operator? These requirements should be detailed in the safety requirements specification for the functions concerned. It might also be appropriate to specify the state related control functions should adopt; you might want a signal to the BPCS to place loops in manual with a fixed default output – this would not be part of the SIL rated provision, but it might avoid operational embarrassment on start up when the protection is reset. It may be appropriate to arrange for a reset to be via a limited duration pulse signal to reduce the potential for a faulty reset command being continuously applied.

In the thrust and parry of process plant operation, there is the danger that the temporary, if it works, becomes permanent. If you have declared a function to be SIL rated, you have declared it to be deserving of more than the common or garden degree of rigour in its management. Which is why you should avoid declaring provisions to be SIL rated 'just to be safe': if the functions are not perceived as being 'special' you will undermine the credibility of the truly special.

## 41. CONSEQUENCE MITIGATION IN LOPA

There is a difficulty with LOPA when a mitigation layer does not eliminate the postulated consequences for the scenario but only reduces them. So, for example, if the mitigation was effective, we might postulate a single casualty instead of ten. But in conventional LOPA, the layers are all assumed to act to stop the realisation of the postulated consequences against which a tolerable event frequency has been assigned. How then may we proceed?

An approach that has been advocated is to have one LOPA study with a mitigation PFD and the full consequence, and another without a mitigation PFD and therefore increased frequency but with reduced (mitigated) consequence (and increased tolerability). Depending on the tolerable frequency - consequence calibration being used, these two LOPA may yield different results if being used for SIL determination purposes.

Alternatively, we may aggregate the mitigated and unmitigated versions of the scenario as follows:

Without mitigation we might identify a risk reduction requirement for our preventative SIF of 500. If the mitigation was always effective in reducing the consequences, the increased tolerable event frequency might mean that the required risk reduction for prevention was only 50. If the mitigation was effective 9 times out of 10, we may identify the aggregate risk reduction requirement as the average from 1 at 500 and 9 off at 50; an average of 95 across all 10 times. The implication is that the risk reduction requirement reduces from 500 to 95; a factor 5.26 change corresponding with an effective additional PFD contribution due to mitigation of 0.19

From these considerations, if we identify the actual probability of failure on demand of a mitigation defence as $PFD_M$ and that layer as providing a Mitigation Factor of MF; being the factor by which the tolerable frequency increases for the reduced consequence, we may identify the effective PFD (the reciprocal of the **additional** risk reduction factor) for that layer as:

$$\frac{\textit{Required RRF with mitigation}}{\textit{Required RRF with no mitigation}} - \frac{\left\{RRF_{NM} + \frac{RRF_{NM}}{MF} \cdot \left(\frac{1}{PFD_M} - 1\right)\right\}}{RRF_{NM}}$$

The numerator here shows the smaller required RRF as being the average requirement due to the Mitigation Factor being applied n-1 out of n times when the mitigation is successful, with n being $1/PFD_M$. This, thankfully, reduces to

$$PFD_{Eff} = PFD_M + \frac{1}{MF} \cdot (1 - PFD_M)$$

and is independent of the required risk reduction. This $PFD_{Eff}$ may then be assigned to the mitigation defence layer within a **single** LOPA in the normal way.

To expand on my example, let us assume we have a fire or toxic release suppression system that would reduce the number of casualties if it were successfully deployed. We postulate that without mitigation there would be 10 casualties. With effective mitigation we anticipate 1 casualty.

If the tolerable event frequency for 10 casualties was $10^{-6}$ per year, and that for 1 casualty was $10^{-5}$ per year, the mitigation factor (MF) would be $10^{-5}/10^{-6} = 10$. If we have a probability of failure on demand for the suppression system ($PFD_M$) of 0.1, the $PFD_{Eff}$ for the layer (from the above equation) would be 0.19

So, in the LOPA study, when we identify the mitigation layer, we would assign it an effective PFD of 0.19 rather than the 'raw' 0.1 associated with the probability of failure of the mitigation defence. The tolerable frequency assigned would be that for the full (unmitigated) consequence.

Given the shift from 0.1 to 0.19 it can be immediately seen that the aggregate LOPA claims less from the mitigation layer. It may be considered a more faithful modelling of the risk associated with the scenario.[a] I don't say this approach is necessarily to be preferred over the 'two LOPA' approach; as always, the tool should be employed with a proper understanding of the underpinning foundations and the fact that the result remains highly uncertain.

The approach may effectively capture some nuances such as a conditional modifier for immediate ignition which would then give rise to a pool fire or, if ignition was delayed, a flash fire or explosion.

An indicative range of values for $PFD_{Eff}$ is shown in the table:

| $PFD_M$ | Mitigation Factor (MF) | | | | |
|---|---|---|---|---|---|
| | 2 | 5 | 10 | 20 | 50 |
| 0.5 | 0.75 | 0.60 | 0.55 | 0.53 | 0.51 |
| 0.2 | 0.60 | 0.36 | 0.28 | 0.24 | 0.22 |
| 0.1 | 0.55 | 0.28 | 0.19 | 0.15 | 0.12 |
| 0.05 | 0.53 | 0.24 | 0.15 | 0.10 | 0.07 |
| 0.01 | 0.51 | 0.21 | 0.11 | 0.06 | 0.03 |

**Table 41.1** Indicative values of $PFD_{Eff}$.

If tolerable frequency calibration is proportional to consequence, (so, for example, no scale aversion is employed) then the order in which mitigation layers are invoked would not matter and the approach may be used in more than one layer in a given LOPA.

---

[a] SISSuite allows the specification of MF in LOPA conditional modifiers.

## 42. SIL4

SIL4. First thought; wow! Second thought; really? A SIL4 project is *'not to be undertaken lightly, but advisedly and soberly'* (I borrow here from the marriage service). And as with marriage, anyone embarking on such an enterprise should do so with a full appreciation of what they are getting into. You need to be certain 'it is the real thing'. In that respect it is worth checking:

a) that the tolerable risk target is not unrealistically ambitious
b) that the SIL determination exercise did not employ unduly conservative assumptions or estimates
c) whether the process design is flawed and could be usefully revisited to enhance the inherent safety aspects and reduce the dependence on active SIF protection

(I.e. Whether there is any just cause or impediment…)

In fact, both 61508-1 (7.6.2.11) and 61511-1 (9.2.11) require reconsideration of the application to see if it can be modified to avoid SIL4 and point to the possibility of using multiple protection layers, including additional **independent** SIF (see Chapter 38). It is perhaps not so very challenging to meet the architecture and systematic capability requirements by using provisions that are both redundant AND diverse; the requirements might then be met by a configuration that would otherwise be limited to SIL3. Depending on the specifics (element type, diagnostic coverage etc.) there may be a requirement for additional hardware fault tolerance, so 1oo3 or 2oo4 say, instead of 1oo2 or 2oo3, but this is merely a matter of extrapolating familiar instrumentation considerations to support the enhanced integrity requirements. The real challenge with SIL4 comes with the 'softer' aspects of compliance: verification, assessment, auditing, validation, planning, project management. The degree of rigour required in executing these aspects, together with the commensurate degree of independence for verifiers/assessors/auditors, makes for a very much more difficult project.

I postulate a SIL4 project would typically need an order magnitude increase in man-hours. That is not to say that the project timescale would increase correspondingly of course; by introducing more resources the timescale could be compressed but no amount of resource would see it scaled back by the postulated factor ten. I can well imagine the aggregate of the 'SIF' bars on the Gantt chart growing by a factor of at least two.

The additional demands on demonstrated rigour and quality assurance may be familiar in the culture prevailing in the nuclear or pharmaceutical sectors but would be alien elsewhere and correspondingly challenging. Even in the nuclear power sector, SIL4 is unusual; the approach is rather to pursue inherent safety and multiple layers of protection.

SIL4 territory is so obscure that I have avoided specific guidance within this book, although many of the considerations outlined will still be relevant of course.

# ERRATA

Any errata will be published via the SISSuite website:

www.sissuite.com

# INDEX

## A

activation energy, 120, 121
aggregate risk, 34, 167, 168
ALARP, 1, 2, 35, 36, 49, 50, 87, 91
Arrhenius, 119, 120
authority, 61, 62, 67, 116, 156, 159, 160

## B

B10, 121, 122
benefit, 2, 13, 35, 52
beta factor, 205-207, 217
BPCS, 38, 91, 92
Buncefield, 52

## C

C&E, 28-30, 47
calibration, 14, 42, 50, 78, 92, 134, 149, 150, 152, 153
certificate, 17, 18, 65, 104, 109, 111
certification, 17, 64, 65, 75, 105, 109
*Challenger*, 4, 7
change control, 61, 88
commissioning, 24, 96, 97
common cause, 200, 201, 204-207, 217, 218
common mode, 112, 140-143, 149
Common Mode, viii, 204

compound SIF, 172
conditional modifiers, 38, 41, 77, 85, 192
confidence level, 16, 135
conservatism, 16, 21, 37, 50, 51-53, 85, 131
consultants, 22
contractors, 22
coverage, 59, 61, 78, 93, 101, 141, 142, 144-147, 150, 153, 154, 158
CPD, 51, 57
cybersecurity, xi, 195, 196

## D

decommissioning, 190
demand, 9, 15, 16, 40, 41, 42, 47, 77, 82, 100, 102, 114, 122, 136, 137, 140, 141, 144, 145, 146, 151, 156
diagnostic coverage, 206, 231
diverse, 175, 208, 213, 218, 231

## E

Engineering Council, 14, 19, 51
EUC, 130
expert, 2, 55, 159, 161

## F

failure count, 115, 119, 123
FARADIP, 67
Feynman, 5, 7
firmware, 66, 80, 117

fitness-for-purpose, 10, 83
FMEA, 17, 67, 105
FPL, 72, 73
FSA4, viii, 177, 178, 181, 236
FSA5, viii, 189, 190
FVL, 72, 73

# G

GAMBICA, 26

# H

Hallberg-Peck, 119, 120
hardwired, 221
HazOp, 21, 46, 48
HEART, 223
HFT, 9
HSE, 34, 49, 53, 90, 159
human error, 40, 76, 220, 223
hypothetical person, 167, 168

# I

individual risk, 34, 42
inspector, 93, 159, 160, 161, 162
integrators, 22

# J

jeopardy, 46, 48

# K

Kipling, 163

# L

Lagging indicators, 177
Leading indicators, 177
legacy, 13, 64, 81, 82, 84, 86, 89
lifecycle, 18, 20, 24, 60, 63, 90, 95, 98, 99, 145
LOPA, vii, ix, x, 8, 37, 38, 41, 52, 170, 171, 175, 176, 184, 216, 228-230, 236
LVL, 72, 73

# M

Management of Change, 226
mandatory, 54, 97
mission time, 102
Mitigation, viii, ix, 183, 185, 186, 229, 230
modification, 6, 61-63, 73, 75, 76, 86, 97, 104, 148, 190
MRT, 102
MTBF, 102, 103, 105, 109, 111, 112, 123
MTTF, 102
MTTR, 102

# N

Normalisation of deviance, 4
nuclear, 221, 232

# O

occupancy, 38, 41, 43
overhaul, 93, 124, 147, 150
override, 61, 148, 180, 192, 224, 225, 227

# P

Pangloss, 53
partial coverage, 93, 158
periodic, 57, 63, 92, 145, 150
Poisson, 68, 69, 178-181, 236
prior-use, 17, 27, 64, 65, 80

procurement, 24, 27, 66, 82, 86
professional judgement, 2, 12, 14, 16, 18, 19, 51, 161
proof test, 31, 32, 61, 63, 69, 78, 87, 88, 92, 94, 102, 105, 144-147, 151, 152, 156, 158, 189, 206, 210-212
proportionality, 49
proven-in-use, 10, 27, 64, 65

**Q**

qualitative, 8, 49, 79
quality assurance, 62, 80, 82, 100
quantitative, 8, 49, 79

**R**

random hardware failure, 17, 75, 77, 78, 82, 105, 116
repair, 61-63, 77, 79, 88, 102, 142
rigour, 2, 9, 14, 15, 16, 18, 49, 51, 70, 71, 75, 79, 86, 96, 99, 106, 132, 136, 145, 151

**S**

Security, viii, 60, 62, 166, 195, 198, 199, 203, 226
Security Levels, 198
SFF, 9, 69, 70, 71, 115, 142
SINTEF, 75, 83, 105, 113
SRS, 86, 95, 97, 130
suspended load, 191, 192
Swiss cheese, 6, 191

systematic capability, 10, 77, 79-81, 106
systematic failures, 10, 15, 75-79, 82, 105, 157

**T**

THERP, 223
tight shut-off, 86
tolerable risk, 8, 34, 36, 37, 85
tolerance, 16, 69, 83, 134, 135, 151
trip setting, 86, 130-132, 134, 136, 151

**U**

Uncertainty, 14, 15, 133
useful life, 62, 88, 101, 102, 116, 119, 120-124

**V**

validation, 24, 76, 96-98, 145
verification, 24, 76, 86, 89, 90, 92, 95, 96, 98
vintage, 124
vulnerability, 51, 106, 107, 109, 110-113, 116, 117

**W**

warranty, 67, 121
waterfall, 20, 164

**Z**

zones & conduits, 198
Zoning, 39

# SISSuite™

**SIS**Suite™ software offers an integrated suite of modules for the management of functional safety throughout the safety lifecycle. The modules are fully linked for appropriate data transfer. The suite includes the following features:

- Hazop Module
- Functional Safety Planning Module
- Version control for all critical records with traceability of originator/checker/approver.
- Cloning of records to simplify data entry
- Management of access for record check/approval through a user competency register
- Advanced LOPA that distinguishes between Pre & Post SIF layers and accommodates mitigation layers.
- Process Safety Time calculation module that links to the SRS module
- Automatic inflation of failure rates for equipment operating beyond its nominal useful life
- Flagging of equipment that is operating beyond its nominal useful life
- Calculation of prevailing PFD on the basis of age inflated failure rates
- Facility for identifying element failure rates on the basis of vendor data/user data/generic data (with approximate Poisson distribution calculation for user data)
- Full SIL compliance verification; PFD, HFT (route $1_H$ or $2_H$) and systematic capability.
- Facility for periodically validating nominated failure rates during the operating phase of the safety lifecycle (in support of FSA4)
- Facility for leading indicator measures during the operating phase of the safety lifecycle (in support of FSA4)

## ABOUT THE AUTHOR

Harvey T. Dearden BSc CEng FIET FInstMC FIMechE FIChemE has worked in the process industry sector for over 40 years and has been employed by vendors, contractors, consultancies and end users. Since 2003 he has managed his own consulting practice. He is known to many through his involvement with the professional engineering institutions and his numerous papers and articles.

He is an InstMC Registered Functional Safety Engineer and an IChemE Professional Process Safety Engineer.

Printed in Great Britain
by Amazon